Apology to Kelp

The authors contributed equally to the creation of this book.

ISBN 979-8-9965536-0-0
Library of Congress Control Number: 2026915057

Names: Written by Becs Epstein & Max Bogan
Title: *Apology to Kelp*

Shivering Blue Press, Chicago, IL
Second Edition

www.becsepstein.com

Max Bogan (ORCID ID: 0000-0003-0270-8590)

The book cover is designed by Becs Epstein.

Table of Contents

Some Logistical Matters

Tragically but inherently, communication always carries the risk of misunderstanding and misinterpretation between parties. In this book, we use a variety of tools/devices as communication aids - sometimes to draw attention, sometimes to distinguish between authors, and sometimes to make abstract or undefined concepts more concrete. We summarize these devices below to facilitate a clearer understanding between us, as authors, and you, the reader.

Green type:
The making of this book involved crosstalk between concepts prevalent in the Arts and Sciences. Both of these fields often rely heavily on specialized language. To clarify the meaning and connotation of words that we considered 'jargon,' we define jargon-y terms in green type. See page 19 for further context.

Burnt orange and purple type:
The majority of text in this book is presented in black text, and these portions were written based on substantive contributions from both Becs and Max. However, each of us had ideas and assertions that arose entirely from our own experiences. We use purple text to indicate portions of the book written entirely or almost entirely by Max; we use burnt orange text to indicate portions of the book written entirely or almost entirely by Becs.

Kelp pronouns (they/it/you):

In the plural, we use ‘kelp’ to refer to *a collection* of organisms that all fall into the same evolutionary group. E.g. “Kelp are large, fast-growing seaweeds.”

In the singular, we use ‘kelp’ to refer to *the category* of organisms that fall into the evolutionary group Laminariales. E.g. “... kelp is used as the scenery to central narratives...”

We also use ‘kelp’ in the singular to refer to specific individuals. E.g. “Image 8. Max holding a bull kelp.”

I grew up amongst the towering redwoods and the vibrant California poppies. I know at some deep space inside of me what it means to be kin with the forest; I know what it means to feel safe against the rough bark and how it feels to have an innate protectiveness of these woods. In this book, we invite you to imagine, with us, the ocean as home. Remember what home means to you and dive with us beneath the canopy of the kelp forest.

- Becs

Preface: Kelp 101

Picture a forest.

Now picture it underwater.

Think of dappled canopies stretching toward sunlight; but instead of birds flitting between tree trunks, imagine otters and fish swimming through slender green and brown stalks that extend 40 meters high.

This is a kelp forest. Can you sense home here?

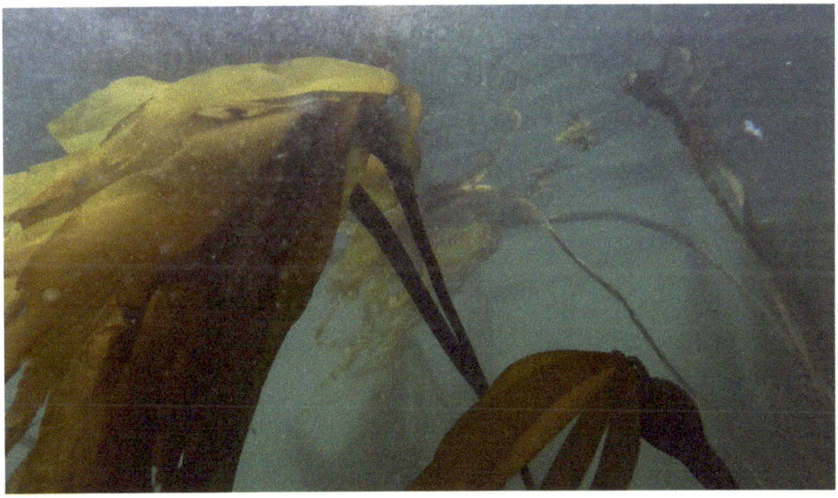

Image 1. Bull kelp *(Nereocystis luetkeana)* near the shore of San Juan Island.

Kelp are large, fast-growing seaweeds. They are members of the brown algae, a group of eukaryotic organisms so evolutionarily ancient that the most recent shared ancestor between plants and brown algae existed more than a billion years ago (Yoon et al. 2004). Within brown algae, kelp make up the order Laminariales ("true kelps"). Their closest relatives include rockweed and maiden's hair. Though we compare brown algae to terrestrial plants, keep in mind that these organisms have fundamentally different cellular and biochemical properties.

Trees are a part of our human and literary history from *The Giving Tree* by Shel Silverstein to *The Overstory* by Richard Powers. Yet in the popular culture of the United States (US), kelp is more likely to be used as the scenery than in central narratives, such as in the Disney animated *The Little Mermaid*. However, kelp is a staple resource and companion in a similar way to trees. Kelp provides food, shelter, and nurseries to numerous other organisms. Kelp forests prevent erosion and stabilize their environment - though unlike trees, kelp do this by slowing rapid ocean currents. And like botanical life, kelp are major nutrient cyclers that shape the chemical landscape of their ecosystem and provide a base to the food chain.

Much like the old-growth forests of the Pacific Northwest United States, or the Amazon forest in Brazil, kelp forests have experienced profound population loss. Global estimates suggest that 38% of kelp forests have experienced

Eukaryote: one of the broadest evolutionary groupings. Generally, eukaryotes are defined as organisms whose cells contain a nucleus, or a separate compartment for DNA. **Clade:** an evolutionary grouping that includes a common ancestor and all the descendants (living and extinct) of that ancestor (University of California at Berkeley, Museum of Paleontology). **Heterokont:** also called stramenopiles - a major eukaryotic taxonomic group including diatoms and brown algae. **Seaweed:** seaweed is a non-scientific term used to refer to almost any macroscopic algae - this can include red algae, green algae, or brown algae. Famous seaweeds include nori (genera *Porphyra* or *Pyropia*), sugar kelp (genus *Saccharina*), and bull kelp (genus *Nereocystis*). A shared descriptor of 'seaweed' does NOT mean that organisms called seaweeds are closely related to each other. Also, a tasty snack.

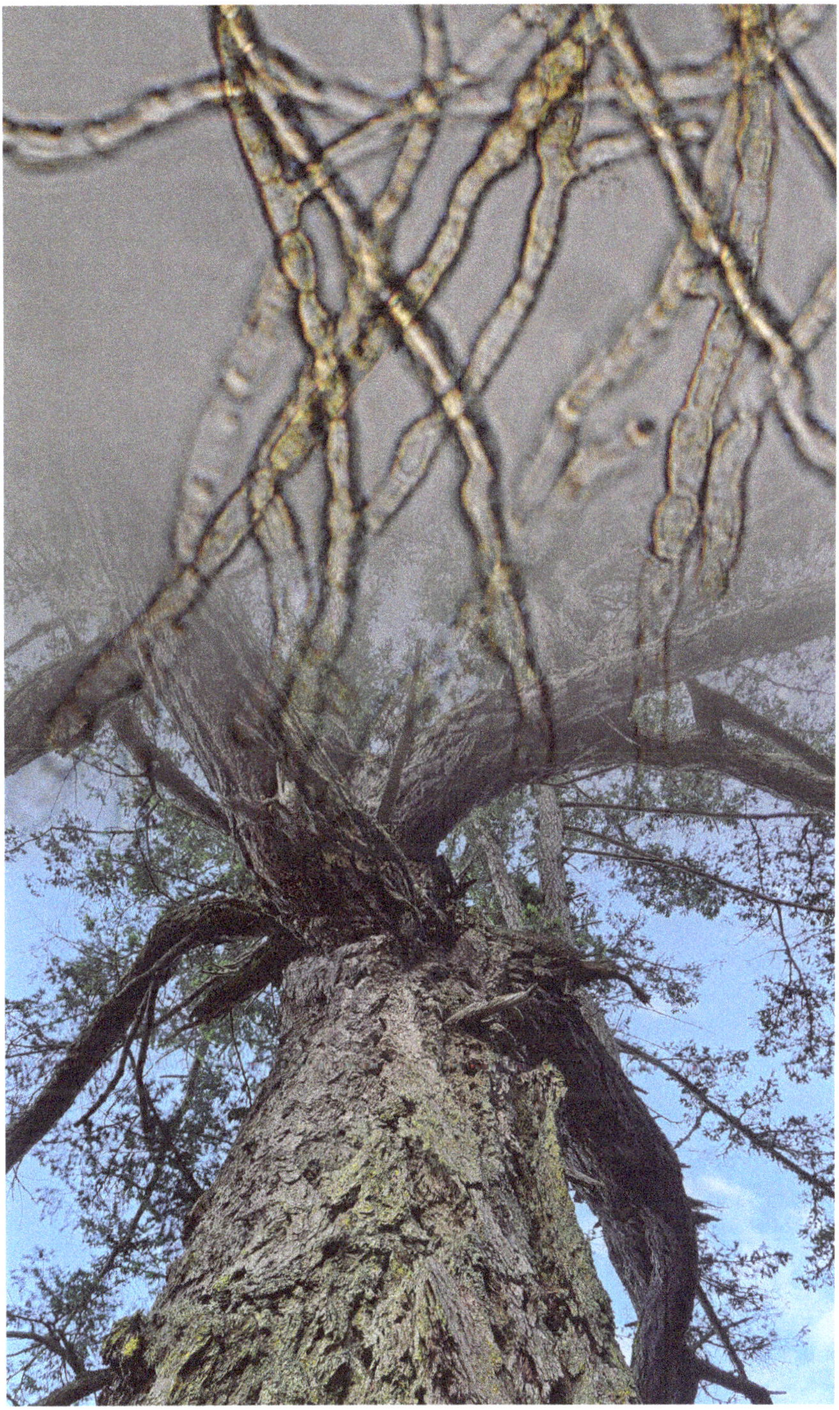

Image 2. Compilation of images of juvenile bull kelp taken at University of Chicago and a tree on San Juan Island.

decline, most within the last 10 years (Krumhansl et al. 2016). Along the Pacific coast of the US, estimates of canopy loss range from ~63% in the Salish Sea (Berry et al. 2021) to over 90% along the northern coast of California (Rogers-Bennett and Catton 2019).

We can smell the smoke, as fire rips through the forests and we run our hands through the ash. As humans, we sense danger when the birds no longer sing, but down in the kelp forest what can we see when kelp is decimated? We can mourn kelp on its own, but we must also mourn the host of organisms that make kelp their home.

Kelp forests are home to thousands of creatures. Look for the long stalks with bulbous appendages. See a school of fish move between the blades; a crab rests on a pneumatocyst. Try this out as your new home for a minute. Home here means a place for existence as well as a place of symbiotic care - kelp offers safety and refuge to their intertidal cohabitants, while drawing carbon dioxide from the atmosphere to the benefit of organisms as remote and far-flung as us humans.

The drivers of kelp forest loss are complex. Some are abiotic: higher ocean temperatures and acidification will increasingly drive disease and mortality in multiple kelp species. Kelp population loss is also caused by interactions between kelp and other organisms. Like the chestnut trees across the United States, huge numbers of sunflower sea stars have died out due to a wasting disease that spread in 2013.

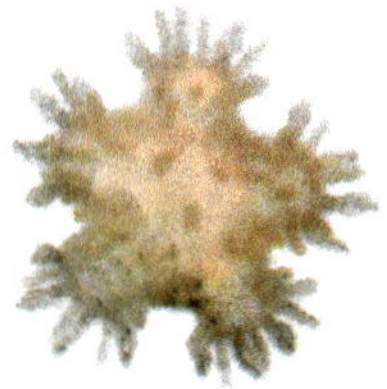

Image 3. Juvenile sunflower sea star.

Abiotic: describing chemical or physical forces that are not considered living. E.g. temperature, light intensity, or concentration of some molecule. **Cosmovision:** the understanding that non-humans like watersheds, forests, and mountains, have personhood. **String Figures/SF** (as used in *Staying with the Trouble* by Donna Haraway): the space in between, science fiction, speculative feminism, science fantasy, speculative fabulation, science fact, string figure, about ethics and knowing stable ground is already troubled and intertwined.

Sea stars' prey, purple sea urchins, are major consumers of kelp, and since 2013 have experienced explosive population increases. The results are catastrophic, with swathes of kelp forests replaced by "urchin barrens" that blanket the seabed of the intercoast.

The balance is broken within these kelp forests. The stalks are gone and what is left is urchins as far as the eye can see. Let the original kelp forest be your home as you flip through this book. But always, always, at the periphery, is the reality of the devastated landscape beneath the waves.

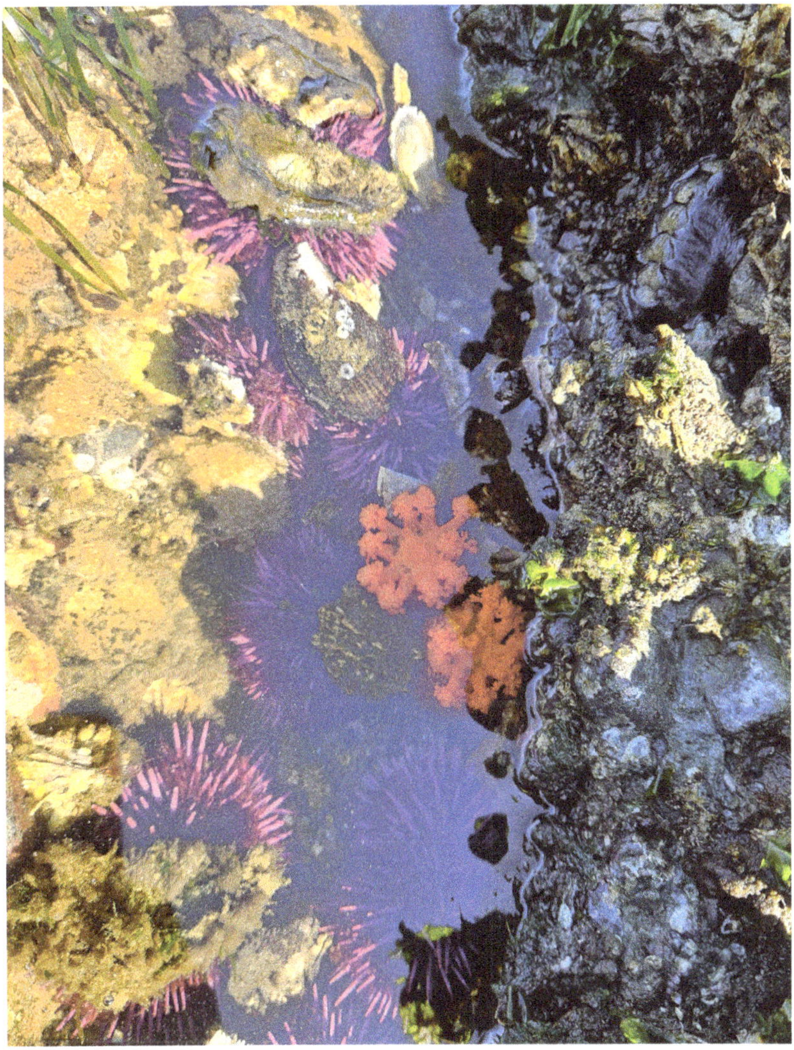

Image 4. Intertidal assemblage, image from Kydaka Point in Washington.

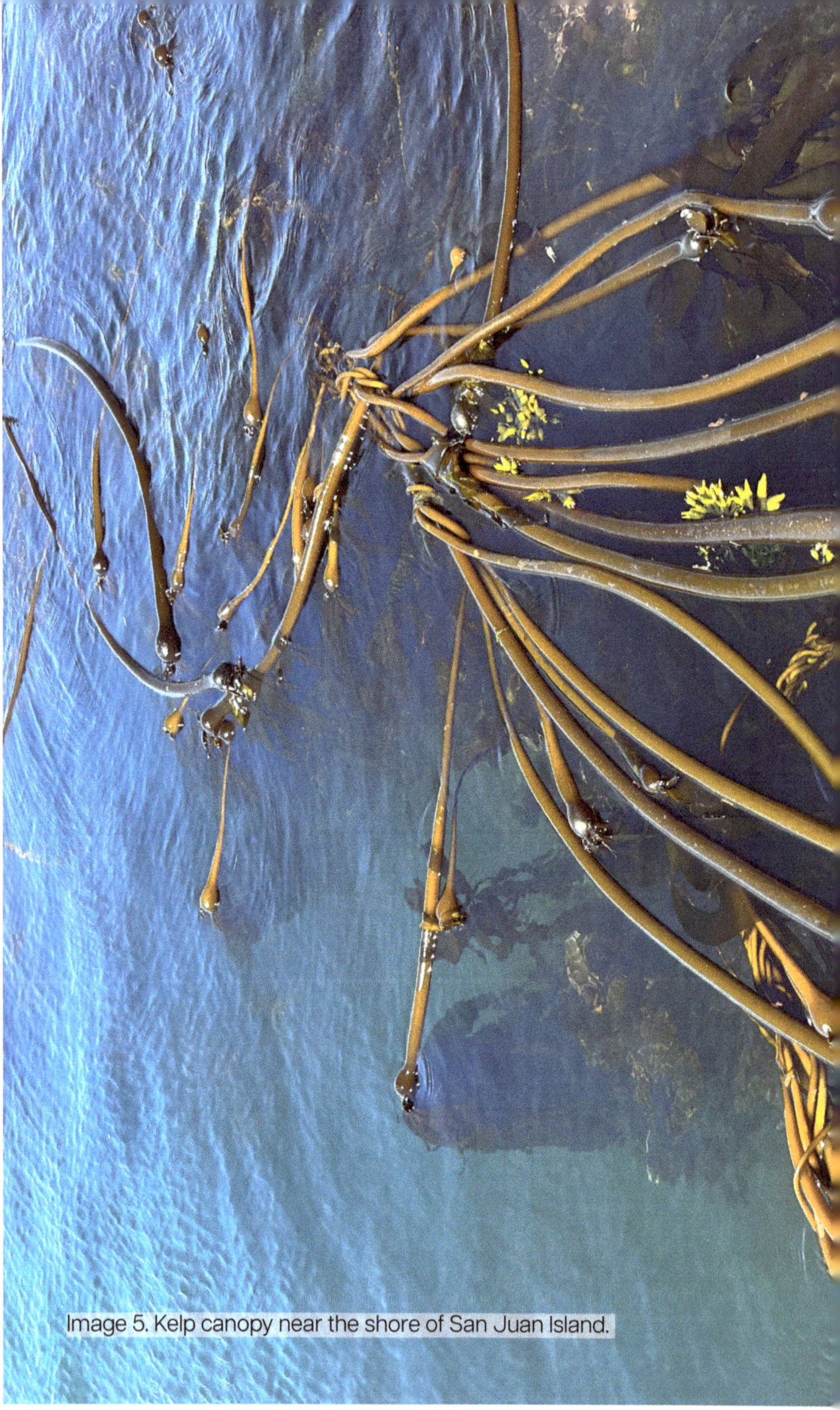

Image 5. Kelp canopy near the shore of San Juan Island.

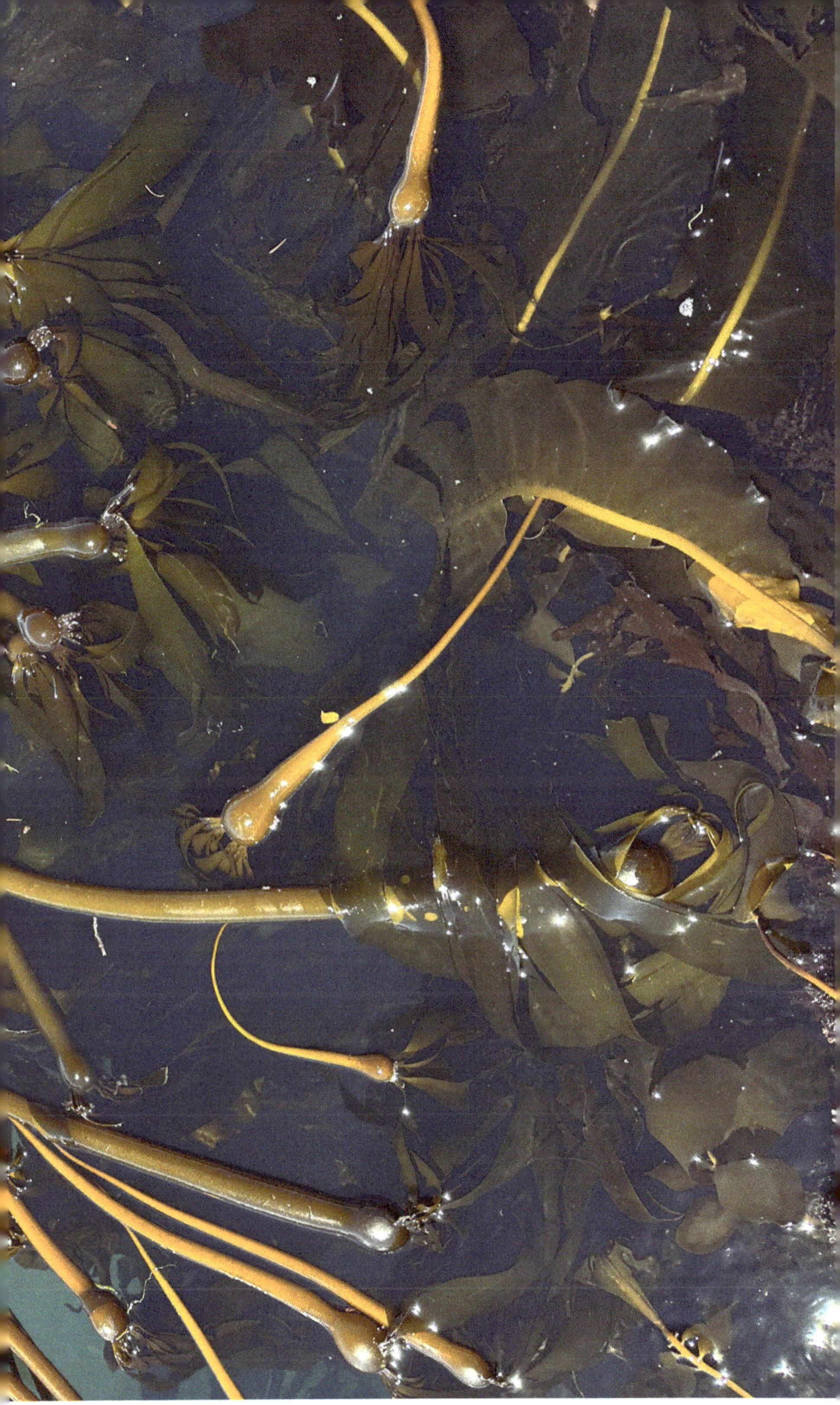

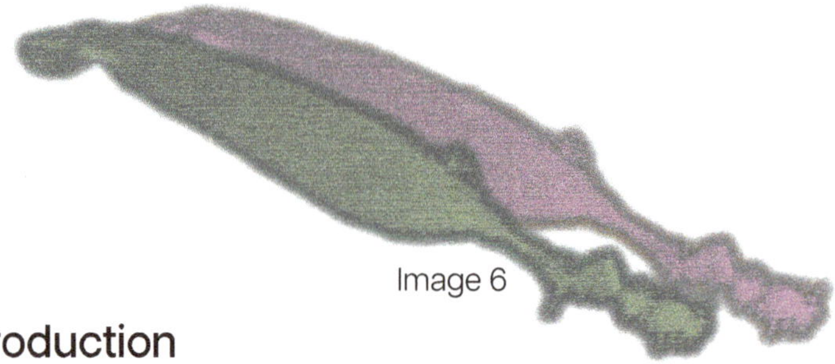

Image 6

Introduction

Apology to Kelp is an interdisciplinary project that says "sorry" to kelp. As foundation species in the intertidal ecosystem, kelps engage in inter-organism interactions throughout their life cycles. The interactions between kelp and humans have been extremely imbalanced, and harmful to kelp: while humans benefit from ecosystem services and commercial resources extracted from kelp, kelp have experienced the dramatic consequences of anthropogenic climate change and corresponding trophic cascades. Our attempt to apologize to kelp stems from a desire to make amends, to reconsider the human relationship to kelp.

This is a project from Becs Epstein and Max Bogan in which they apologize to kelp from scientific and artistic points of view. This book documents how our apologies, and our research, intertwine and diverge. *Apology to Kelp* includes collective writing, external submissions, and kelp experimental homemaking.

Becs' practice explores the dynamics and subtleties of human-plant relationships through research-based ceramics, performance, painting, and installation. By blending handcraft and digital technology, she creates work that is assertive yet

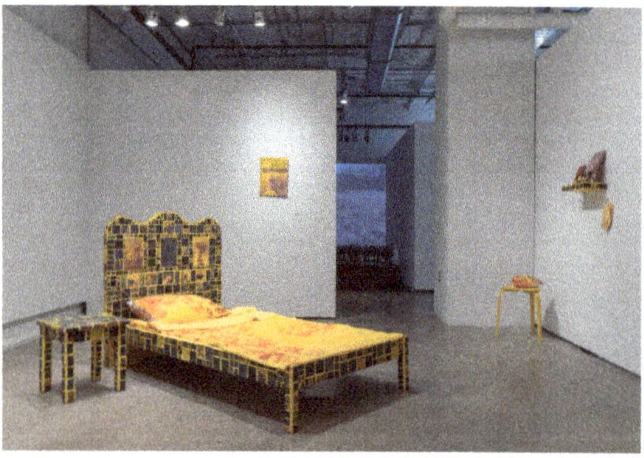

Image 7. Becs' thesis art.

dream-like, navigating the intersection of anthropocentrism and ecocentrism in the context of the climate crisis. A recent year-long project involved apologizing to a different plant each day through speech, sculpture, ceramic, and written text. While inherently insufficient, these apologies represent hopeful failures – opportunities to look at utopia from the other side. She confronts humanity's ecological impact and experiments with fostering coequal kinship with botanical life. She received an MFA from School of the Art Institute of Chicago in 2024 and a BFA from Carnegie Mellon University in 2017.

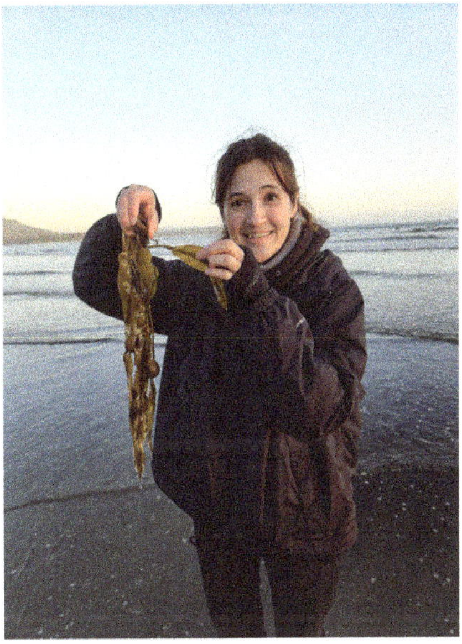

Image 8. Max holding a bull kelp.

Max is a PhD candidate in evolutionary biology at the University of Chicago. Max's work attempts to understand kelp from molecular, ecological, and evolutionary perspectives. The organisms that gave rise to modern day kelp emerged tens of millions of years ago. However, we still have precious little information about how the unique interactions between kelp and their ecosystems have evolved in that time.

Anthropogenic/Anthropocene: anthropogenic describes the quality of being attributable to human activity, especially in the context of environmental impacts. The Anthropocene is a new geological era, wherein the climate and environment of the planet is so heavily influenced by human action that it is no longer continuous with the previous geological era, the Holocene. **Biochemistry:** explains biological characteristics as the product of the fundamental properties of matter and energy. **Evolutionary biology:** a field of biology that is primarily concerned with questions of how things came to be as they are. Often described as a 'historical' science, evolutionary biology accounts for the characteristics of living systems in terms of their descent and divergence from a common ancestor over time.

Through her research on how kelp interact with their environment, Max hopes to learn about these organisms at the level of their biochemistry and DNA. By apologizing to kelp concomitantly with that research, she reconsiders the arbitrary lines we draw between ourselves and other organisms. Based on her education as a biologist, and her upbringing in a biracial family, Max finds these lines arbitrary and often actively harmful. Biological diversity exists on a continuum, and the consequences of global environmental change cannot be cleanly partitioned into 'effects on humans' and 'effects on non-humans.' As we seek to understand kelp and other non-human organisms in the context of our own harmful impacts, we must address the framework of dominance and exploitation that often dictates the ways that humans relate to and study the non-human.

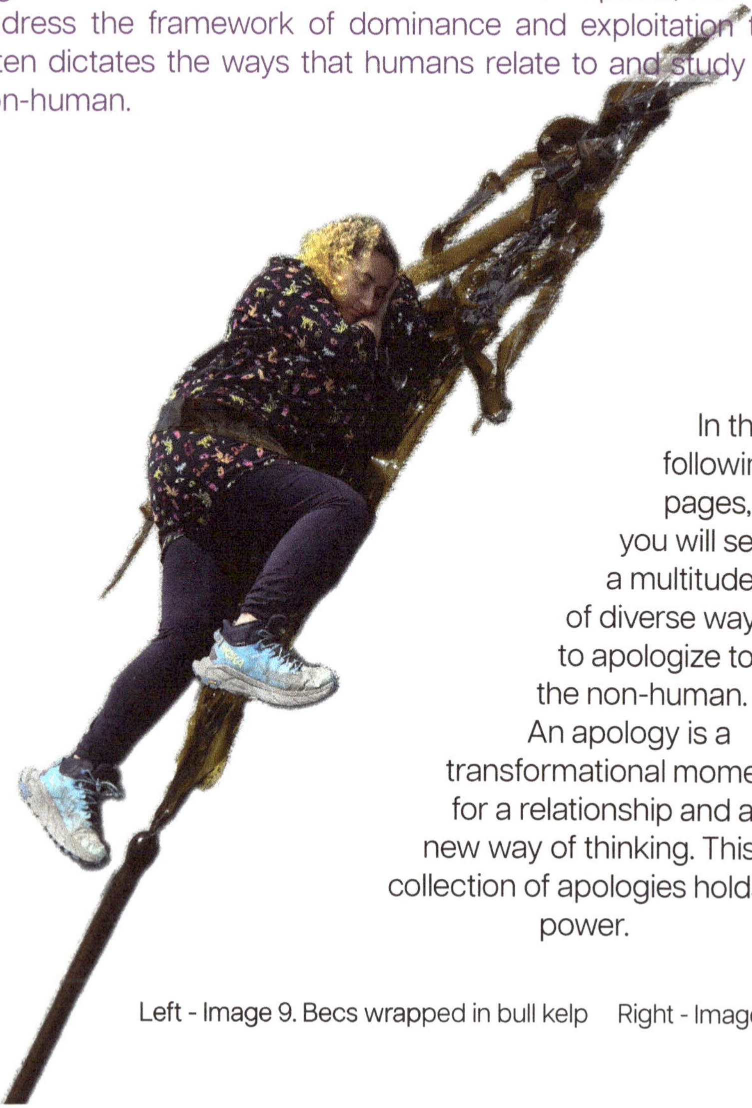

In the following pages, you will see a multitude of diverse ways to apologize to the non-human. An apology is a transformational moment for a relationship and a new way of thinking. This collection of apologies holds power.

Left - Image 9. Becs wrapped in bull kelp Right - Image 10

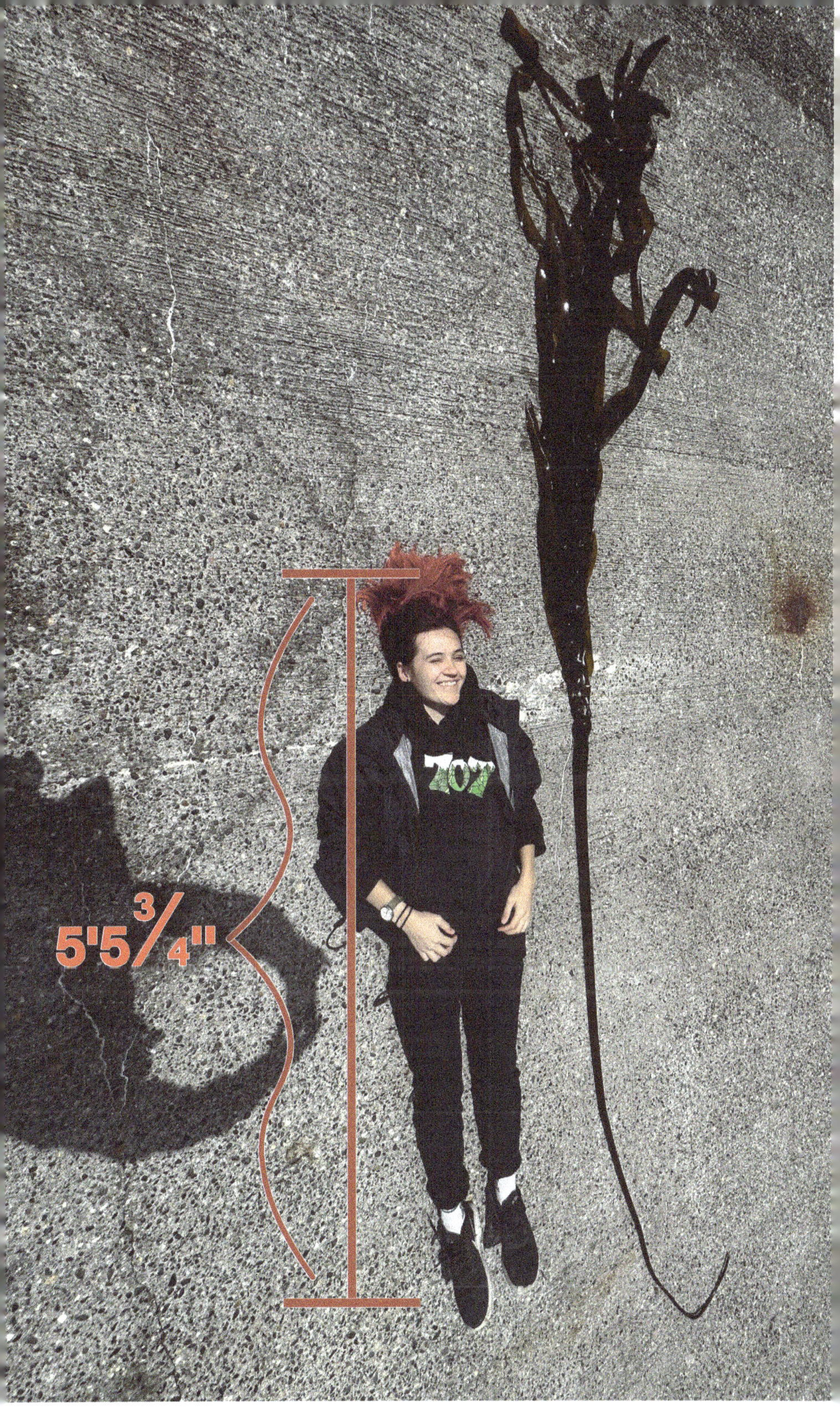
5'5 3/4"
707

Role of Science & Art in Apologies to Kelp

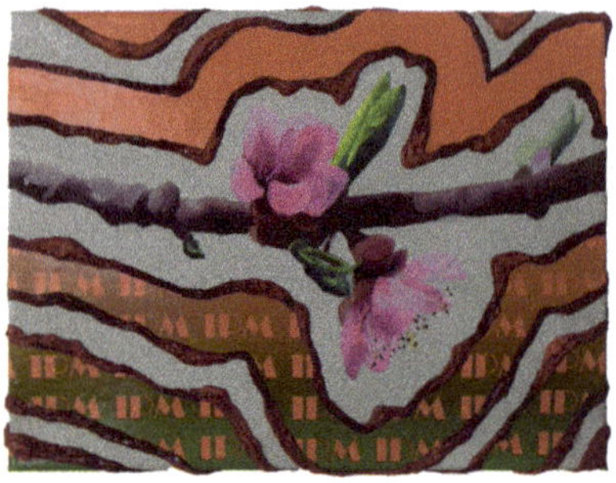

Left - Image 11. *Blossoming Moment* by Becs

Right - Image 12. Dissecting scope

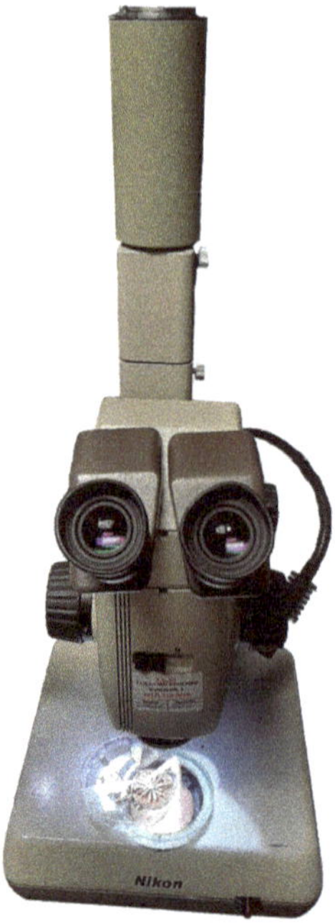

Art and science are often viewed societally as opposite disciplines, but in practice they converge and diverge in complex ways. Both fields ask questions about our interpretation of reality, and often require creative methods to find probable answers.

Art defies definition, but to attempt the impossible: art is a way to understand, a source of beauty, a perspective, and an emotional epiphany. Art can be motivated by personal experience, aesthetic sensibilities, and societal issues. While art can take endless forms, a work of art is always defined by three essential criteria: meaning, embodiment, and interpretation by the viewer (Arthur C. Danto, *What Art Is*). Art is fickle in that it can hide in the shadows or yell from the rooftops, like Mierle Laderman Ukeles shaking every sanitation worker's hand or Christo and Jeanne-Claude wrapping L'Arc de Triomphe

Translation: an interpretation of the world or of information. Different translations may not directly correspond to each other, or engage with each others' conclusions. The process of applying the results of basic scientific research to address a particular biomedical problem; the process of expressing ideas articulated in one language in another language.

entirely in fabric. Art can be heavily researched or intuitive or anywhere in between. Art is, in its greatest form, a way to communicate, but it is not always comprehensible to everyone.

Scientific inquiry is the practice of explaining the world based on observations. Sometimes our observations appear contradictory, incomplete, or simply baffling. We might seek to explain such observations, or to predict what we will observe in other contexts. Sometimes observation is the goal in itself. In any case, the first step of scientific research is to determine what is known – what has been seen, measured, identified, and ideally validated by multiple independent sources. Science connects us to the world around us, and helps humans understand how to care for non-humans. But it is not always a welcoming field. Both the process of science and the reporting of scientific findings can be extremely insular.

Combining the languages of science and art overcomes barriers to fruitful communication with broad audiences. Collaboration with hard science gives the art a solidity behind the morphing sandstorm of artistic thought. In turn, collaboration with the arts provides scientists access to a novel set of tools and media to express their conclusions and convictions. Practice in both art and science teaches an important skillset; these disciplines train practioners to identify the connections between things, and articulate why they are important. In *Apology to Kelp*, both artistic and scientific insights are crucial for articulating the relationship between humans and kelp. Max and Becs are finding the overlaps and missed connections within their respective disciplines in their apologies to kelp.

As part of this merging of fields, they started to build a personal glossary. You will find this dispersed at the bottom of pages in green type. Whenever Max and Becs used incomprehensible jargon or even just a word in an unfamiliar

Formalizing: In terms of art, this means giving structure to the art to give it power. For example, it was formalizing when Becs created a system to apologize every day for a year to a plant. She created a calendar for each month to track and archive these apologies. Becs first tried to explain this by saying emphasis on the "alizing." **Poiesis:** making and sustaining. Sympoiesis for making with others, autopoiesis making itself.

context, they created their own common language by writing down definitions. Most of these are not dictionary definitions, but they are how Max and Becs understand meaning and explain it to each other. The traveling glossary melds their academic and lived experiences to help each other, and the reader, understand. Becs and Max often wrote collectively or back and forth, but when they write entirely separately the text is coded in purple (Max) or burnt orange (Becs).

What is an Apology

What makes an apology? What makes it meaningful? In the course of writing this book, Max and Becs sought definitions of apology from each other, and from friends, advisors, and colleagues.

Max and Becs define apology slightly differently, but believe the core of the act is trying to say "I'm sorry" and to express remorse. Becs believes that an apology is meant to admit harm and be an attempt to be better. Apologizing accepts the possibility of failure while hoping for an eventual win. Win is defined here as receiving forgiveness. The goal is to really mean IT and to acknowledge the harm you have done while accepting that you may not be forgiven. It is a potential transformational moment for a relationship. Max believes that an apology requires relinquishing power, and offers the harmed party a path forward, toward, or through. To apologize means an attempt to give up or turn over the power to harm, and to acknowledge that having that power in the first place imbues you with responsibility. Apology should be paired with a commitment to change or stop the behavior that caused the harm, though the degree of commitment may vary.

Max and Becs asked others their thoughts on apology:

Sara Black is an artist who uses processes of building and horticulture, as well as materials like diseased wood, inherited building supplies, and ecosystem-specific plants, to reveal how people and things are entangled in shared worlds. Her work has been exhibited internationally. She is an Associate Professor of Sculpture at SAIC and a fellow at the Neubauer Collegium.

Sara believes that an apology evokes an act of reconciliation, that needs something beyond feeling remorse. It represents a marriage of sorts: it says, I have committed to care about you, to be concerned about you. In the case of apologizing to kelp, it is an exercise in both hope and loss.

Josie Iselin is a photographer, author, and designer of many books, with her last few focusing on the world of seaweeds and kelp. She is also the co-director of Above/Below, a campaign working to bring the recognition afforded the forests on land to the kelp forests below the ocean's surface.

Josie believes apology means that you are listening: even between humans, it is difficult to get out of our own stories, and compassionately imagine the story from the perspective of the person you are apologizing to. If an apology cannot be accepted, then there is nowhere to go in the relationship. To apologize to kelp is therefore a start, though we should look to non-colonial cultures and ask how they apologize to the non-human.

Sean Hoban is the Senior Scientist in Tree Conservation Biology at The Morton Arboretum, where he has worked since 2016. He is a faculty member at the University of Chicago. His team aims to produce the knowledge and tools to inform conservation action and policy, focusing on conserving and restoring populations and their genetic diversity.

Sean thinks that apology involves recognizing harm done, admitting your role in it, and expressing regret and empathy. It may or may not involve attempts to repair damage.

Apology has countless definitions, making it an ideal way to relate to the non-human. The common thread of the definitions collected above suggests that apologies should be moments of humility, and of empathy. To apologize to kelp in this spirit, Max and Becs have drawn on the philosophies of both science and art. The tools of scientific inquiry help them understand and articulate the magnitude of the harm they and others do to kelp and kelp forests. Knowledge of the ecology and evolution of kelp allows Max and Becs to identify possible remedial actions that aspire to reduce the ways in which kelp are damaged, stressed, and discarded every day.

The practice of art demands empathetic reflection from Becs and Max and offers these two humans a conceptual way forward, even as they know their apology may never be received or accepted. Kelp can't read a letter or look into a human's eyes as they say sorry, but art can translate between legible words and non-human communication. An apology to kelp is ephemeral in that from mouths beneath the water, the sincerity bubbles around the stalks. There is no scientific answer to why their apology is warranted. It is impossible to understand the degree to which Max and Becs have personally negatively impacted kelp - it borders on arrogance to believe that two individuals could do anything substantive to something as foundational and powerful as kelp forests. But it is possible and necessary to reflect on what behaviors of theirs contribute to the harm kelp experiences when situated in the context of rising ocean temperatures, increased chemical pollutants, and other forms of global change.

Human & Non-human Relationships

As we discussed in the previous section, apology is a fundamentally relational act. It requires honoring your connection to the apologizee. An apology is an act of humility and attempted empathy that can momentarily subvert hierarchy and power dynamics between two entities. To contextualize our apologies to kelp, we invite you to dive into Posthuman thinking with us.

In the United States, starting often in kindergarten, we are taught that humans are the top of the food chain. We are shown animals as classroom pets and plants as semester long projects. The guinea pig is often over or underfed and the sunflower sprout slowly suffocates as it outgrows its paper cup or withers from lack of water. This mistreatment and occasional death is, at best, utilized as a teaching moment or, at its worst, we learn to accept negligence towards non-humans as the teacher moves onto the next lesson. The non-human is presented as an object that we can fully control. This is all a long way to reference the common Western relationship with the non-human: humanism. Apologizing to kelp is a task shakily set within Posthumanism. Before we can converse on why we want to work in a posthumanist mindset, let's dig into defining these two concepts.

Posthumanism decenters the human as the only changemaker on the planet. Posthuman theory suggests that the hierarchical dynamic that has structured human-to-non-human relationships throughout much of the last five centuries is not (and maybe never was) sustainable. Francesca Ferrando, a leading voice in Posthuman Studies and founder of the NY

Posthuman Research Group, wrote

> "Posthumanism offers a theoretical invitation to think inclusively, in a genealogical relocation of humanity within universality ('Posthumanism' as a criticism of humanism, anthropocentrism and universe-centrism), and alterity within the self ('Posthuman Ism' as a recognition of those aspects which are constitutively human, and still, beyond human comprehension)."

Ferrando is putting forward the idea that Posthumanism opens up an entryway to placing ourselves within a broader ecological perspective where humans are not the exception.

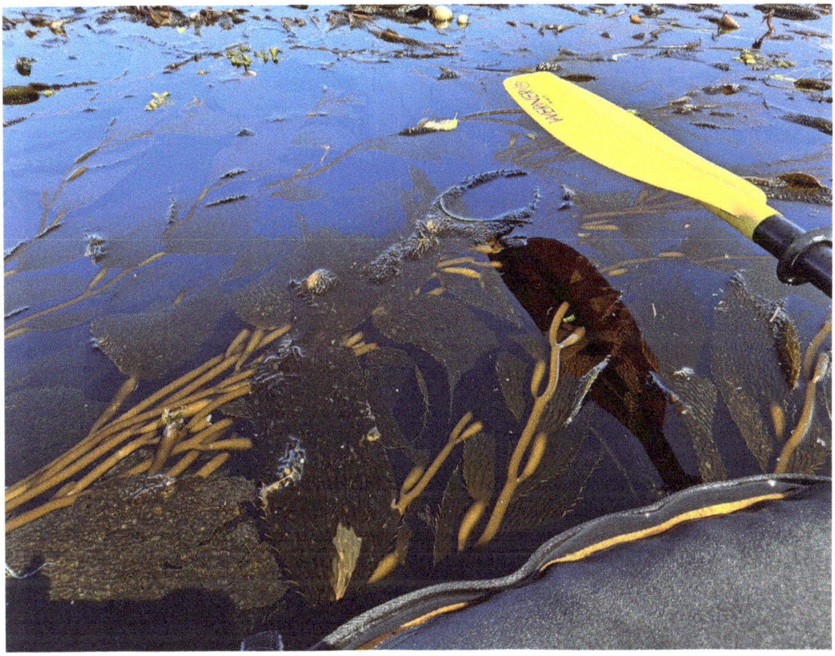

Image 13. Macrocystis blades floating near Bullman Beach, Washington

Humanists see many characteristics as purely present in humans but Ferrando writes that while these aspects

Kincentric: a framework by Enrique Salmon referring to when intrinsic value images from humans practicing a relational, familylike connection with the life and systems of life that make up the biodiversity and cultural complexity of a place. This all feels aligned with Salmon's ideas of becoming keystone species again.

exist within humans, that does not mean they are absent in other organisms. This idea pairs with these characteristics being essential elements of humans while still not being fully understandable to humans. Posthumanism is undoing the separation of human-self and object, an idea introduced to the public imagination by John Locke.

Robin Wall Kimmerer (Potawatomi botanist, author, and the director of the Center for Native Peoples and the Environment at the State University of New York College of Environmental Science and Forestry) wrote in "Maple Nation" in *Braiding Sweetgrass* of maple trees as citizens, going on to say:

> "Drinking maple syrup gives you quite a sugar rush. This tool is what it means to be a citizen of Maple Nation, having maple in your bloodstream, maple in your bones. We are what we eat, and with every golden spoonful maple carbon becomes human carbon. Our traditional thinking had it right: maples are people, people are maples." (Kimmerer 2013, 172).

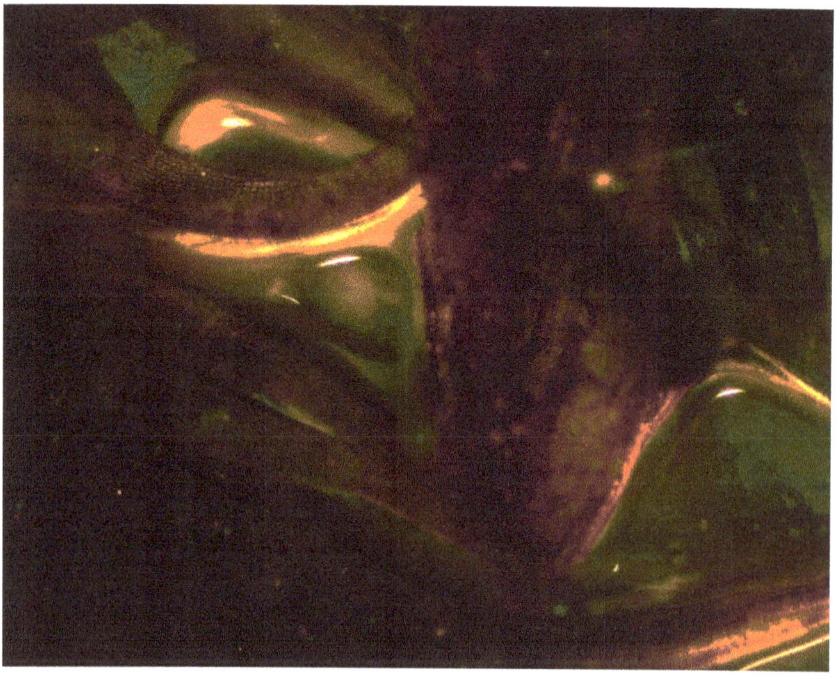

Image 14. Microscopy image of various moss taken by Becs

In this reading, maple trees have as much of a right to representation as any human. The maple exists not just in symbiotic partnership with humans, but in this case, as kin with each other. The dynamic becomes muscle memory and an ancestral heritage. This is a posthumanist viewpoint where no organisms are the central figure. Being in community with these trees means becoming sticky with them. This means relationality not in a hierarchy, but in a cat's cradle as described by Donna Haraway, a prominent scholar in the field of science and technology studies. In *Staying with the Trouble*, Haraway proposes creatures' interconnectedness by way of the String Figures of a cat's cradle where response-ability is cultivated. She criticizes anthropocentrism while offering her own posthumanist and ecofeminist framework.

Sam Frost, a political scientist, writes of humans as biocultural creatures – humans are shaped by their past and present biology and environment. This is at a literal corporeal level as well as within our intangible thoughts. This complicates what it means to be an individual. Anna Tsing in *Unruly Edges: Mushrooms as Companion Species* writes that

> "Human exceptionalism blinds us...Human nature is an interspecies relationship. Far from challenging genetics, an interspecies frame for our species opens possibilities for biological as well as cultural research trajectories."

She leads us through understanding entanglements through mushrooms as nourishment of body and mind. Her writing is sticky with tendrils of Posthumanism. Fungi show us that we are not alone as a species; we are entangled and colleagues. The maple syrup is part of our bodies in equal measure to being part of our spirit. These thinkers and writers begin in Anthropology, Political Science, and Critical Plant Studies, yet they all bring different angles to how Posthumanism

Chthonic/Chthulu: We are all interconnected and the ending isn't determined. We have hope. Donna Haraway said in *Staying with the Trouble*, "made up of ongoing multispecies stories and practices of becoming-with in times that remain at stake, in precarious times, in which the world is not finished and the sky has not fallen – yet" is not only true, but also that thinking in a Posthumanist way enhances our way of life.

On the other hand, humanism has been the primary relational framework for most of western and colonized society for hundreds of years. In this section, we argue that humanism has harmed our current existence and extensively damaged humanity's potential ecological future. John Locke wrote in *The Second Treatise on Government* [1689] that: 'Every Man has a Property in his own Person. This no Body has any right to but himself' (II § 27). Ed Cohen breaks that sentence down to its core in *A Body Worth Having? Or, A System of Natural Governance* by paraphrasing it as "It seems a simple enough formula: To be a (legal) person means to have a body." Following the thread of western political theory, 'the body' and 'the soul' which once were conjoined ideas were cut from each other and forced into the shape of individual beings. Locke put into being the concept of inherent power within each human, and in turn turned the physical body into a vessel with firm boundaries that can be defended. The separation of body and soul was a tool to forcefully pull power from the monarch.

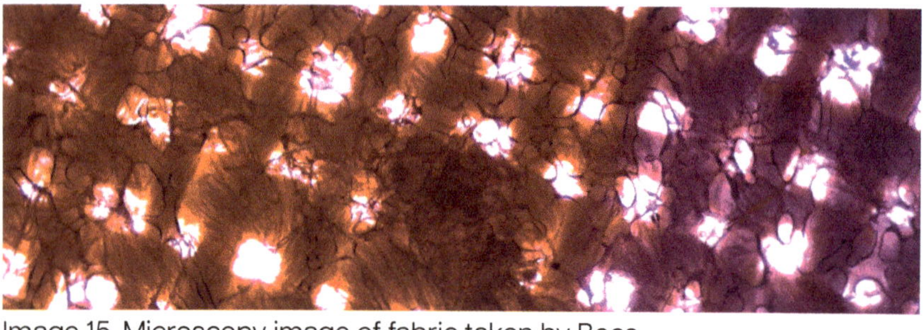

Image 15. Microscopy image of fabric taken by Becs

In England, the monarch ruling based on divine right began to chafe on citizens' understanding of self. John Hobbes in *Leviathan* (1651) presented a case for political sovereignty not based on divine authority. This enabled the public to move power away from royalty. The passing of the *Habeas Corpus Act* of 1679, built on that framework, still asserted the monarch as sovereign yet it constrained that power as having to answer

Frame: (as defined by Lakoff 2010) a mental structure that includes semantic roles, relations between roles, and relations to other frames. E.g. a "hospital" frame includes roles like "Doctor," "Nurse," "Patient," and relations between roles such as "doctor treats patient".

to the law of the land.
Cohen goes on to translate *Habeas Corpus* as literally meaning

> "'You may/should have the body for submitting'... whosoever 'has' 'the body' must present 'the body of the party' to the court in order to allow the court to exercise its legal jurisdiction, which is in fact the jurisdiction of 'the law' per se."

This 'body' becomes legally a limit to the violent power of the crown. At this point, you may be wondering why this chapter is diving deeply into the political upheaval of the 1600s, so let's take a step back. The threads that weave together Humanism braided themselves thicker and more all encompassing during that time period.

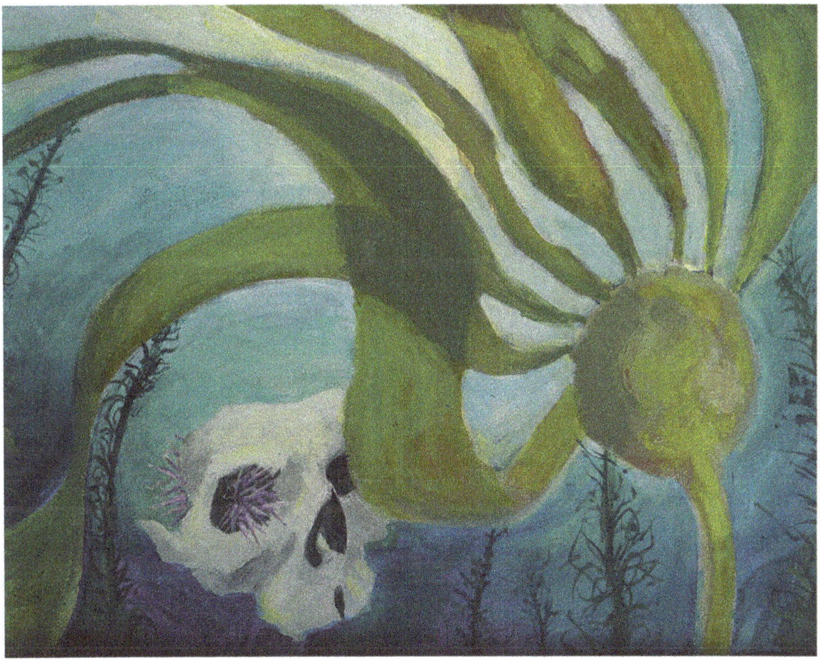

Image 16. *Momento mori* by Max

In the process of fighting against unjust rulers, thinkers of the 1600s and 1700s created new political imaginings. These novel theories of Hobbes and Locke dig into ideas of what makes a human reliant on themself as opposed to on the whims of a divine ruler. That humanist braid wraps around the ancestors of current western thought and to this day leaches toxins into our collective consciousness. Claiming the inherent rights of 'the body' is the way in which a subject could stake its formal claim to being a citizen and thus entitled to the due process of law.

'The body' here acts as a legal fiction as strong as the body of the monarch. Locke goes as far as to clearly place 'the body' and personhood as a social and political issue and not as an emotion based locality.

> "'For should the soul of a prince, carrying with it the consciousness of the prince's past life, enter and inform the body of a cobbler...the soul, with all its princely thoughts about it, would not make another man: but he would be the same cobbler to every one besides himself"

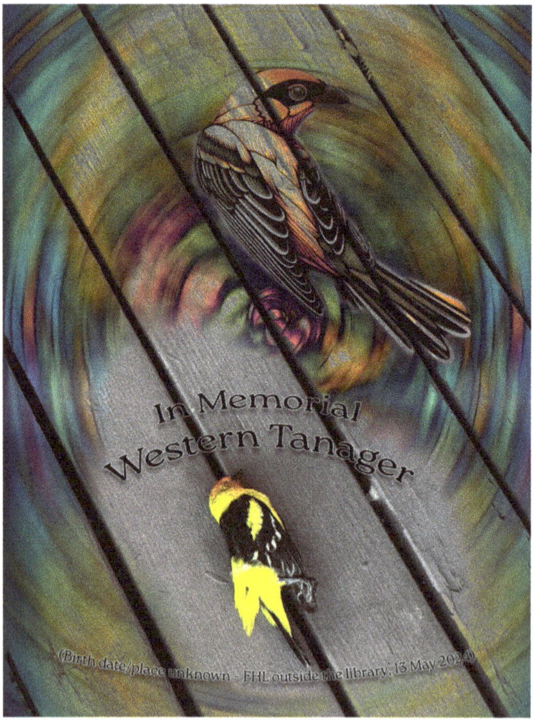

Image 17. Created by Becs for a bird she found dead at FHL

The soul here is a personal inner feeling while the visible body is the acknowledged self in public space. Humanism comes out in full force here by treating only the physical body as holding the right to protection and respect. We must ask what becomes of the non-human under this strict definition of power. Humanism centers and elevates the human experience and Lockian thought defines that humanity with strict guidelines. This leaves no room for roots of the non-human to wiggle into power. How can kelp be protected when the only legal beings end at the human epidermis? Humanism is inherently a hierarchy which places humans as the ruling class; a hauntingly similar premise to Locke's intention, just with a twist.

Finally, we argue that reevaluating our relationships with non-humans under a posthuman framework mutually benefits humans and non-humans. bell hooks has argued eloquently that the oppressive system of patriarchy injures both the men that it elevates and the non-men that it subjugates. While individual women, trans, and gender nonconforming people experience direct injury through material discrimination, individual men are injured through the expectation (often but not always from other men) that they perform maleness in a way that upholds patriarchal power (hooks 2004).

The same might be said of other oppressive systems founded on humanism, including capitalist systems of monopolized resource distribution and consumption. As Anna Tsing has noted, these systems promise power and prosperity to their human followers, while making the resources necessary for stability for all species ever more scarce (Tsing 2015). In *Apology to Kelp*, we reject humanist and capitalist understandings of kelp and other species that are based on the idea of "Pristine Wilderness." This is the concept that the American west was untouched, pure and ready for the taking. This idea was developed by 19th century Romantic writers and artists, and has been critiqued elsewhere as revisionist history that erases and dehumanizes Indigenous peoples and histories (Fletcher et al. 2021; but see also Whyte 2017 and Geib 2022). In reality, the west was changed by Native Americans in deep ecological ways. "The Pristine Myth," as William M. Denevan dubbed it, allowed European settlers to treat the land and its inhabitants as inert resources that could be domesticated, mined, and profited from.

By viewing humans and kelp as co-equal members of the global ecosystem, we try to set new expectations for ourselves. In her essay *Cyborg Manifesto*, Haraway uses the phrase 'women of color' to illustrate how powerful political identities can be forged by allying ourselves across experiences of 'otherness, difference, and specificity' ('Fractured Identities,' Haraway 2006). In this spirit, we recognize the otherness, difference, and specificity of kelp, and seek to ally ourselves across the gaps in our experiences. Our apologies are gestures of a posthuman politics, centered on our desire for social and ecological progress in our relationship with kelp. The apologies require dismantling power systems defined by arbitrary biological boundaries between human and non-human. Reparative solutions to harm in a relationship require stepping outside of the basic assumptions of power within that relationship.

From a humanist perspective, we might argue that kelp has provided and continues to provide a number of "ecosystem services" from which humans benefited immensely. We have discussed the role of kelps as foundation species that contribute to ecological stability. Kelp, and many of their close brown algal relatives, also represent sources of economically valuable chemical byproducts. Algae are ubiquitous in our daily lives. They are present at the molecular level in everything from our food additives to our research reagents.

A posthuman relationship with kelp is more intimate and complex than kelp as "Pristine Wilderness" or kelp as a commercial resource. Posthumanism recognizes its vitality, and its significance across human history. The scientific and social relationship between kelp and humans began long before the current decline of kelp forests, which is due to rising ocean temperatures and associated ecological cascades. We (Becs and Max) are not the first to seek artistic relations and co-equal footing with kelp.

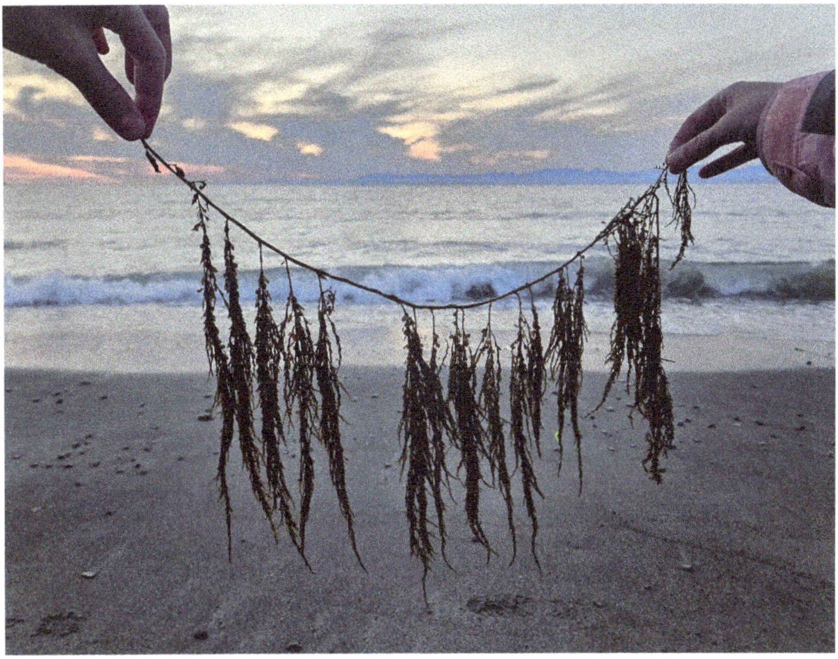

Image 18. Max and Daniela Pierro with algal species at Kydaka Point

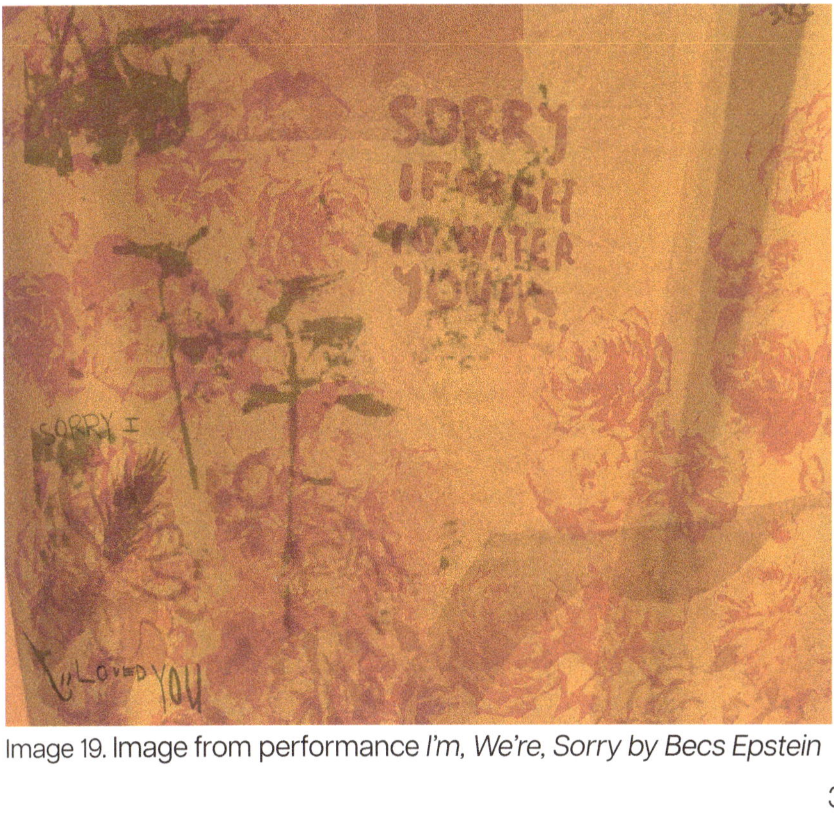

Image 19. Image from performance *I'm, We're, Sorry by Becs Epstein*

The kelp that Max studies (*Nereocystis luetkeana*) features in the cultural histories of a number of Indigenous tribes from the northern Pacific coast of North America. Its roles in these histories are both functional and narrative. Functionally, bull kelp beds have been used as anchors and fishing/hunting grounds by members of the Makah tribe (Naar, n.d.). Narratively, bull kelp has acted as an anchor that connects the human world to the sea. In the early 1900s, the Haida artist Tom Price (born in 1860), and Tom Stevens, chief of Those-born-at-House-point, each told the story of Sinxii' Gangu or Sounding-Gambling-Sticks, a young boy who gambles with a supernatural entity and follows the stipe of a bull kelp into the sea to consult with his grandfather (Glatthaar, 1970; Swanton, 1905). Jaalen Edenshaw, a contemporary Haida artist from the islands of Haida Gwaii (also known as the Queen Charlotte Islands), later depicted the tale in a 2008 play (Davies, 2008; Black Press Media, 2008).

Bull kelp also appears in the Samish people's story of Ko-kwahl-alwoot (KwƏkwálƏlwƏt), a young woman who agrees to marry a sea deity to secure bountiful food for her family. The story goes that, although Ko-kwahl-alwoot left her family to live forever in the sea with her husband, you can still see her hair floating at the surface of coastal waters in the form of bull kelp (Naar, n.d.). The tale, as told by Charlie Edwards in the 1930s and 40s, is shared by the Deception Pass Park Foundation to contextualize a massive sculpture of Ko-kwahl-alwoot made by Tracy Powell. The sculpture shows flowing heads of bull kelp as Ko-kwahl-alwoot's hair. The roles of kelp in Indigenous histories described here are non-exhaustive examples of the shared history between humans and kelp.

Kelp beds more generally are suggested to have facilitated human crossing of the northern Pacific ocean (Bringloe et al. 2020), and the use of kelp in human medical

Homology: is similarity due to shared ancestry between a pair of structures or genes in different taxa. **Taxa:** singular "taxon". A group of biological entities that are categorized according to their shared characteristics. Sometimes used interchangeably with "species," both terms are relative, and imply an intention to compare characteristics between two (or more) biological entities.

contexts continues today. A close relative of bull kelp, Laminaria (species unclear), has been used in gynecological contexts to dilate women's cervices. Dr. Charisse Loder, an OB/GYN at the University of Michigan women's clinic authored a Patient Education pamphlet titled 'Laminaria Placement,' that describes Laminaria as "sterilized, dried sticks of seaweed that absorb fluid from the cervix (the opening of the uterus) and slowly expand to dilate (open) your cervix." The Laminaria is placed before a gynecological procedure like an abortion or birth induction "to gently open your cervix and prevent injury during your procedure" (Loder 2024). These traces of kelp in human history highlight the depth and intimacy of the association between our species.

Raymond Pierotti wrote of the relationship between humans and non-humans as being constantly fluctuating. He quotes Francis LaFlesche, an Osage scholar:

> "The cosmos was in constant motion and consisted of unending, varied cycles of birth, maturity, old age, death, and rebirth. These temporal cycles could not be stopped or reversed, for 'nothing in the cosmos moved backward' (LaFlesche 1995: 30)." (Pierotti 2012, 27).

The world is not meant to be in balance because there are not opposite things to go on the scales. This is a posthumanist viewpoint, to which we can compare humanist understanding of balance. By apologizing, we imagine (or try to remember?) a relationship with kelp that subverts humanist scripts of what we can learn from kelp, of how we should behave in relation to it, and of what constitutes 'balance' in an ecosystem forever changed by human activity.

The conceptual threads of humanism are interwoven with the sociopolitical histories of the United States and many other nations. Our common definition and practice of autonomy is derived from 17th century thinkers who explicitly sought to subvert what they viewed as an unjust hierarchy of power. In this goal, we are aligned. However, where early humanists aimed to empower and protect individuals from the political actions of monarchs, in the 21st century, we aim to protect ourselves as members of a global ecosystem. The elevation of the white, wealthy, male individual has come at the expense of those lacking bodies deemed worthy of legal protection. A posthumanist perspective is integral to this book because an apology inherently places the receiver in an equal plane to the apologizer. A humanist perspective maintains a barrier with humans as more valuable than all other species. A posthuman perspective broadens our viewing field to look at living creatures as interconnected beings who are reliant on each other. By apologizing, we actively choose kinship instead of tyranny.

As an artist and a scientist, our fields of expertise are both heavily influenced by humanist traditions. We asked ourselves, how can scientific and artistic thinking be used to achieve posthuman goals? In what ways are these fields limited by a humanist worldview, and in what ways do they expand our relationships with the non-human?

Max - Why does science need Posthumanism?

To many biologists, the arguments of Posthumanism may seem like obvious extensions of foundational scientific knowledge. Recent discoveries related to the microbiome, as well as long histories of ecological, evolutionary, and cellular research have revealed that most (if not all) organisms depend vitally on bacteria and other microscopic symbionts. Crucial life processes from development to digestion to mood regulation are mediated by the activity of these microbes. It is no longer controversial to state that the biological unit that we observe (e.g. a human) emerges from an assemblage of diverse entities.

This remarkable interdependence is consistent with known principles of ecology and evolution, and is supported by research on organisms across the tree of life. However, Posthumanism offers something that the natural science literature often lacks: Posthuman thinking pairs knowledge of our interdependence with an explicit political framework for how that knowledge should guide our collective behavior. Authors such as Robin Wall Kimmerer, Anna Tsing (an anthropologist and ethnographer interested in human-non-human interactions) and Jane Bennett (an American political theorist and philosopher)have grappled deeply with the consequences of a humanist (or otherwise anthropocentric) worldview under conditions of global environmental change. They recognize human beings as members of an ecosystem of vibrant things. These ecosystems span spatial scales from neighborhoods, to countries, to the whole planet (and beyond?). A Posthuman perspective asks what is lost, when our approach to understanding ecosystems is based on the assumption that humans are separate from and hierarchically related to all the rest of Nature?

In their works, Tsing and Bennett explicitly reference Actor-Network Theory – a theory of thought pioneered by Bruno Latour. I will invoke it here to highlight the relevance of Posthuman thinking to achieving scientifically-supported environmental goals. Actor-Network Theory proposes that 'scientific' and 'social' entities are not truly distinguishable (Latour 2005). Any entity that exists has both 'scientific' and 'social' properties, and entities of interest may be referred to as 'actors' (or 'actants'). As a silly example: a rock is a scientific object whose mass, density, and even history may be measured or inferred; it is also a social object, whose connotations and effects may be observed. The fact that I chose to list mass and density in association with a rock suggests that it is an actor: a rock can act upon me in certain known ways; I have learned that the most important things about a rock are its mass and density.

One outcome of Latour's observation – that the line between 'scientific' and 'social' is blurrier than we might think – is "to present human and non-human actants on a less vertical plane than is common," as Bennett seeks to do in her book, *Vibrant Matter* (Bennett 2010), and as Tsing discusses in depth in her book *The Mushroom at the End of the World* (Tsing 2015). Another outcome is the further erosion of any grounds for the argument that 'the environment' and 'politics' can or should be separated. Bennett asks directly: "How would political responses to public problems change were we to take seriously the vitality of (non-human) bodies?"

When we started writing this book, my relationship with kelp was brand new - less than a year in the making. It was entirely mediated through the practice of western science: I had hypotheses and experimental designs. I was determined to delve deeply into kelp, to pick apart its life threads and reveal the mysteries hidden there.

I still have this instinct, and I think it's an important one. The instinct to understand, to paint as complete a picture of the world as possible, sits at the heart of much foundational ecological research, and it drives fruitful study and knowledge production. Under this model of investigation, humans have learned about kelp across diverse contexts. Using our knowledge of kelp life cycles and developmental requirements, we have learned to cultivate kelp in laboratory and commercial settings. By harnessing molecular and biochemical technology, we have begun to characterize kelp biology at the level of genes and metabolites. Through rigorous sampling and censusing, we have documented the decimation of kelp populations along the western coast of the United States.

And yet, with all the empirical evidence of the 21st century at our fingertips, we still contend with major discrepancies between what we know and what we do, at both individual and governmental levels. At the individual level, we observe an 'attitude-behavior gap' in the United States, wherein many more people in the U.S. say that they would sign a petition, donate time or money, or contact government officials about climate change than actually do (Ballew et al. 2024). At the governmental level, studies from the conservation sciences have found that, across four continents, scientific knowledge is not used effectively to inform conservation policy (Cook and Sgrò 2017; Karam-Gemael et al. 2018). Cook and Sgrò, Karam-Gemael, and a study titled 'Climate Change Engagement of Scientists' (Dablander et al. 2024) appropriately encourage greater involvement, engagement, and outreach by scientists to increase environmentally-aligned policy outcomes. These studies, and the courses of action they recommend, present examples of ways that scientists, as actors with political agency, can honor the vitality of non-human bodies. In *Apology to Kelp*, we aim to take kelp seriously as a sociopolitical entity, as well as a scientific one. When unequivocal, unequal devastation has been dealt to one party by another in a sociopolitical interaction, apology is appropriate. In conjunction with a Posthuman framework, our experiments in apology help us to recognize and honor the vital force of kelp, and imagine them as co-equal actors in our ecosystem.

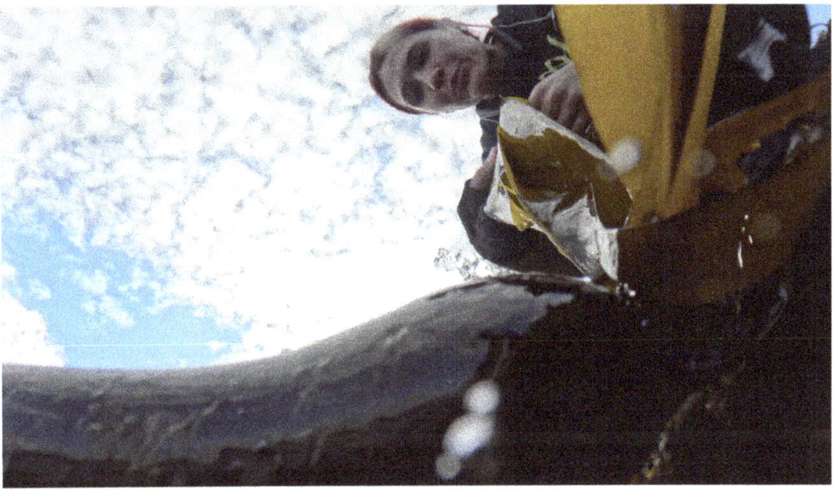

Image 20. Max gargling and spitting on bull kelp to share their mouth's biome with it

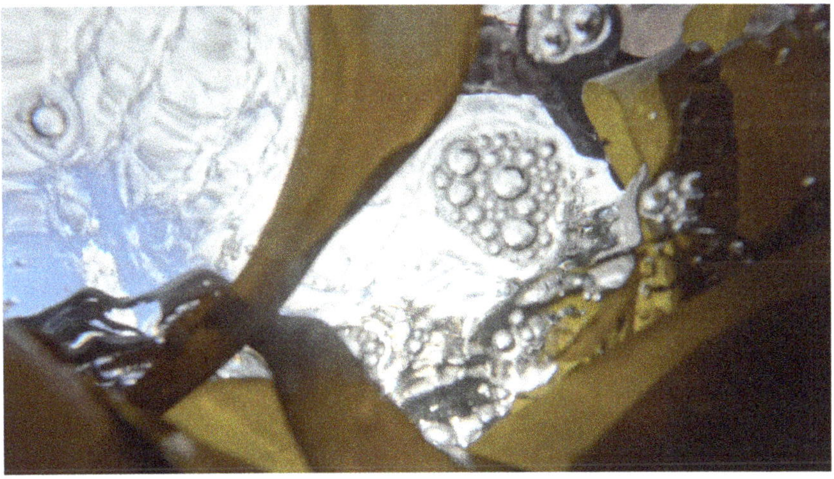

Image 21. After Max spitting on bull kelp to share their mouth's biome

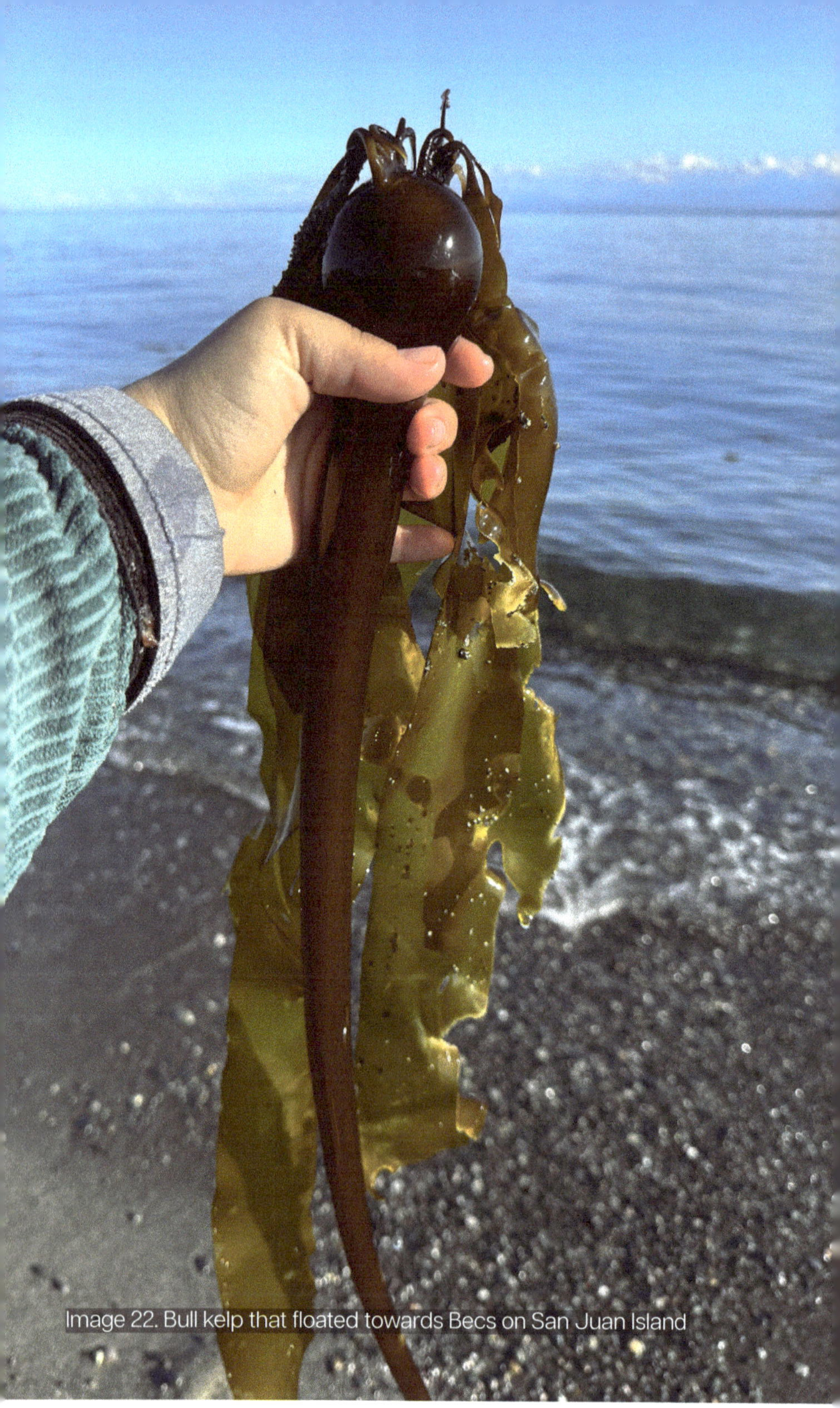

Image 22. Bull kelp that floated towards Becs on San Juan Island

Becs -

Not all art is posthumanist, but the most powerful already is. I hope my own work is posthumanist—an attempt to decenter the human while acknowledging that I inevitably view the world through a human lens. I am drawn to the entanglements between humans and non-humans, the ways in which art can make these relationships visible, legible, and felt.

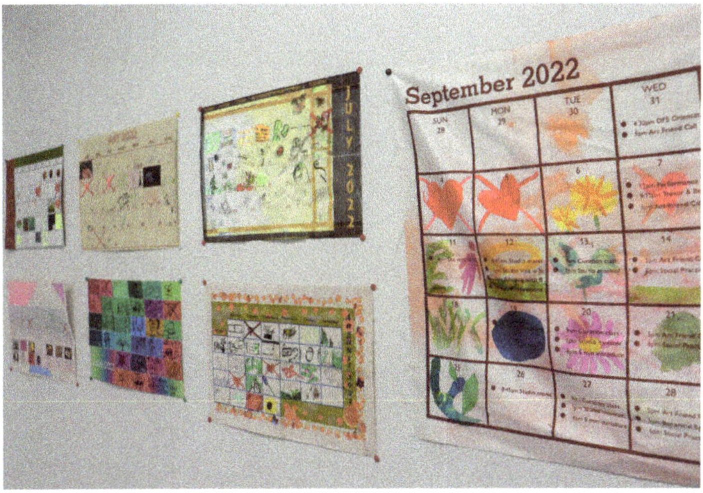

Image 23. *Apology Calendars* by Becs

Posthumanist art resists the idea of human dominance. It challenges anthropocentrism, recognizing that we exist in a web of interdependencies—with soil, with algae, with the detritus of human industry. Artists like Claire Pentecost use soil chromatography to make visible the unseen chemical relationships beneath our feet. Rich Pell's Museum of Postnatural History archives the ways humans have shaped evolution itself. Duke Riley's *DEATH TO THE LIVING, Long Live Trash*, transforms ocean plastic into scrimshaw, reframing waste as artifact, history, and warning. Each of these works offers a perspective in which the human is not the sole protagonist but part of a larger, messier narrative of adaptation, influence, and consequence.

I hope my art, also, operates in this way—the human is not the center but simply one element in a broader system.

Yet, as a human, my understanding is limited. My narratives inevitably reflect back to human experience, not out of choice, but because I am always interpreting through my own body, my own mind. Perhaps this is the paradox of posthumanist art: to strive toward a perspective beyond the human while knowing that the attempt itself is inherently human.

I'd like to step back for a minute and tell you about how Max and I started working together. I proposed the idea of apologizing with Max to kelp, based on my own durational art of apologizing to a different plant everyday for an entire year. I delve more deeply into that foundational art piece later in this book, but let me give you a quick introduction. I attempted to make amends with plant life through a long term reimagining of the dynamic between us. Everyday I would either research a specific plant or come across one in my everyday activities and force myself to consider how society or I had harmed it, or its species. I built up an archive of these apologies which included custom ceramic pots, paintings, performances, and written words. Here are some excerpts from that archive.

4/22/2022
"I put a couple flowering tulip bulbs back into the ground that I found on their side. I was walking to get a drink with my friends like Tiago and Juliana, but suddenly I saw in the planter beds that this whole plant was tipped over. I considered taking it away with me or just keeping on walking. They should not have been planted this early, where did these tulips even get transplanted from that they are so rootbound and weak to the strong Chicago winds. In the end, I stuck my hands in the dirt, made the hole a bit bigger, carefully placed the bulbs back in and gave them more stability by planting them a bit lower. My hands were covered in dirt and the whole group was just sort of looking at me. When I got home that night, there was still dirt underneath my fingernails. I can still feel the cold, cold dirt encasing my hands. I am sorry I did not spend longer with the tulips after I rebuilt their home. I hope they found community with the other tulips around them. I hope no frost touches them and they can live out their days with less human intervention."

09/25/2022
"I said sorry by screenprinting images of my childhood roses on fabric. Today was Rosh Hashanah so it was a celebratory day but also the beginning of a time to think about who I have wronged throughout the past year. I have given many apologies, been thoughtless constantly, but I am hopeful that better relationships are in my future."

I began my year of apologies March 3rd 2022, and seven months later, I reached out to Max on October 12, 2022 including these words: "This must be such a strange email to suddenly get so feel free to ignore it if you aren't interested BUT if you are...I work with human relationships to plants within the dynamic of climate change which is what brought me to stop at your bio." I had been skimming University of Chicago's PhD candidate directory to find someone to apply with for the Arts, Science, and Culture Initiative grant. I looked at anthropologists, economists, chemists, and eventually (obviously) biologists. Max is the first person I messaged. I chose them based on this simple bio "I'm currently interested in studying molecular adaptation to environmental stressors (especially those associated with anthropogenic climate change), with a focus on protein evolution and neofunctionalization." I saw something in that sentence, potentially the "anthropogenic" climate change, which sparked me to want to talk with them. Luckily (magically) for me, Max wrote back quickly and after a video call, we found the fertile overlap between and amongst our interests.

Max and I finished our first draft of this book two months after my official year of apologies came to a close. Before I began apologizing, I already thought of my art as posthumanist, and yet, this experience brought the dynamic between our species into a crystal clear clarity. It was incredibly painful to force myself to stay in the mode of apologizing long term. It was also worth it. Not only did my original project lead to a deepening of my personal relationship with botanical life, but it led me to an entirely new relationship with kelp. I used to think of kelp and algae synonymously as things that tangled around my legs while I tried to swim. Now, I savor the differences between bull kelp (*Nereocystis luetkeana*) and giant kelp (*Macrocystis pyrifera*). I exist in kinship with kelp in that I feel care for them, but also in that Max and I are trying our best to understand them. We try our best to help them while acknowledging that we all need each other - we are in community, not stuck in a cycle of savior complex. That is Posthumanism at its core.

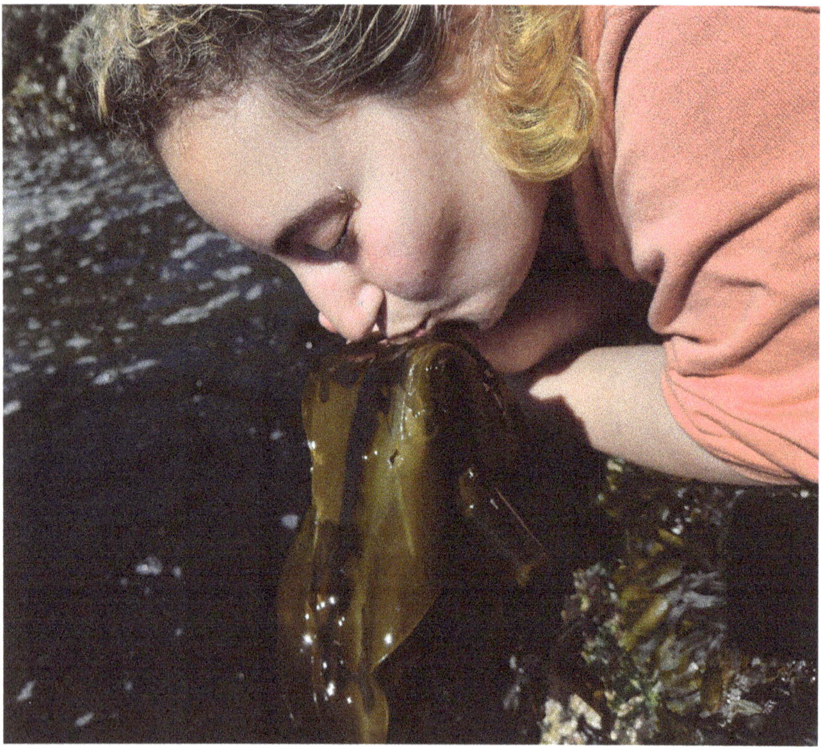

Image 24. Becs kissing bull kelp on San Juan Island

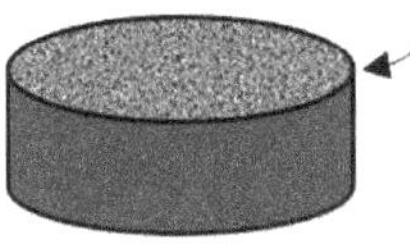

Image 25

Experiment

Motivation

We (Max and Becs) began the aspirational process of reconnecting and remediating our relationships with the non-human by designing an experiment. Our goal was to better understand what conditions in the lab can be modulated to approximate kelp's accustomed home in a complex, turbid environment. In simple terms, we wanted to figure out how to help kelp grow to maturity in a lab setting.

Significant, valiant restoration efforts such as the Puget Sound Restoration Fund are underway to replenish and protect declining kelp populations in Washington state. Conservation geneticists like Felipe Alberto and his lab members at the University of Wisconsin-Milwaukee are making huge contributions to preserve the genetic diversity and adaptive potential of kelps off the coast of California.

Such work tends to the urgent human need to stop the bleeding. From the perspective of ecosystem services, the current state of decline of kelp forests means that humans are hemorrhaging ecological and industrial resources. We (humans) don't know what will happen, or how our lives will be changed if kelp forests continue to suffer and shrink. We (Becs and Max) do know that harm to humanity is not our singular reason for this experiment.

In the spirit of apology, we pursued our shared goals with restoration and conservation efforts, with an intentional focus on heightening kinship and care. The range of light conditions, water chemistry, and temperatures that we can provide in a lab setting are all limited compared to the complexity of kelp's native environment. The burden of adjusting to these differences might represent a significant stressor on lab-grown kelp. It is stressful in that it may harm their growth - but it may also cause sensory pain, or in more anthropomorphic words, it may cause emotional pain. Imagine being removed from your living room which includes a cushy couch and nice temperature and then being put into a totally transformed space that is functional but spartan - a shock to the system, like a couch made of marble.

It is hard to identify good solutions to such stress. We (Max and Becs) are well-meaning but not particularly adept husbands. We attempted to apologize for these shortcomings by creating and testing small ceramic substrates for kelp to grow on in lab settings. We varied the physical and chemical properties of the substrates to see what qualities affected kelp growth. This was both a scientific effort to study kelp more easily in controlled settings, and an emotional homemaking for creatures so far from their wild environments.

For Max, this experiment contributes to apology by creating new relational dynamics with kelp. Max's lab has a particular interest in understanding the interactions between kelp and kelp-associated microbes. Through this experiment, kelp acts as a scientific case study in host-microbe interactions, and as an instructor in how to nurture symbiotic relations, when each partner's language and senses are wildly different. In the end, Max sees each substrate as an attempt to exchange information and small gestures between kelp and humans, as creatures equally able to sense and respond to the cues of our environments.

For Becs, this experiment serves as an introduction, an attempt at making amends through remediating harm, and as proof that science and art combined, have life changing

Grog: high percentage silica and aluminum; first fired, then ground up in to different sizes and incoropräted into a new clay; tends to be porous and of low density. **Substrate:** various meanings - the base on which an organism lives (Merriam Webster) in biochemistry, a substrate could refer to the compound that an enzyme binds to and catalyzes a chemical reaction with. In relation to kelp physiology, a substrate is the surface or object to which a kelp's holdfast (similar to roots) attaches.

potentiality. In contrast to Max, Becs views kelp not as a case study, but as a specific type of living being. Sure, there may be broader scientific and artistic techniques and theories that could extend to other creatures. Kelp is specifically kelp. It is not solely a testing material. From Becs' artistic viewpoint, it feels insulting and diminishing to call such a beautiful and powerful being a case study. This is one of those moments where art theory and scientific research collide. The experience as a non-scientist of watching kelp at different life stages and trying to be supportive but having very little control is like discovering magic in reality.

Methodological Background

Clay, one of Becs' artistic mediums, also exists in this spiritual place of the uncontrollable. It is tangible in its physicality while maintaining a sense of being a living being. It can be broken down into its chemical composition and yet, ceramicists often talk about working with clay and not fighting it; of clay having a mind of its own. This experiment harnesses that vitality by attempting to imbue that feeling to the kelp; and to literally, physically promote the kelp's growth. Sol LeWit, founder of both Minimal and Conceptual art, wrote in *Conceptual Art* (1968) that "Conceptual Artists are mystics rather than rationalists." Artmaking is having faith in not only yourself but in your hands and your subconscious intentions. Clay is shown in this study mostly as stable and quantifiable which is the exact opposite dialogue among ceramists. These two truths of clay can and do exist in this experiment; they are both needed to enable innovative problem solving in science and in art-based apologies. Becs places clay as an opinionated living entity and that treatment shifts to kelp as well. While kelp is obviously alive in a biological way, it is also an elemental force. Kelp elicits respect and even worship, yet as with clay it is with us as our knees dig into the dirt (sand). Like the sun, the oceans, and other elemental forces, kelp and clay are laced into human lives. In this experiment, we tighten those laces and spur all involved parties to be alive in body as well as in spirit.

Growing kelp in a lab is an important way to learn about their life cycles and genetic diversity. It is also especially useful in understanding their microbiomes. Similarly to humans and most other multicellular organisms, kelp host dense communities of bacteria and fungi that live on or within their tissue. Researchers at the Station Biologique de Roscoff have found that brown algae like kelp do not develop normally in the absence of their microbiomes. The idea of providing kelp with a "probiotic" to cultivate a healthy microbiome (think kombucha) has been floated more than once. To study how the addition of microbes alters the growth cycle of juvenile kelp, Max's lab needed a way to encourage maturation, manipulate the conditions of the kelp once they had developed to a visible size, and minimize unnecessary stress.

In the oceans, kelp reproduce by releasing spores into the water column (see Figure 1). These spores settle onto solid

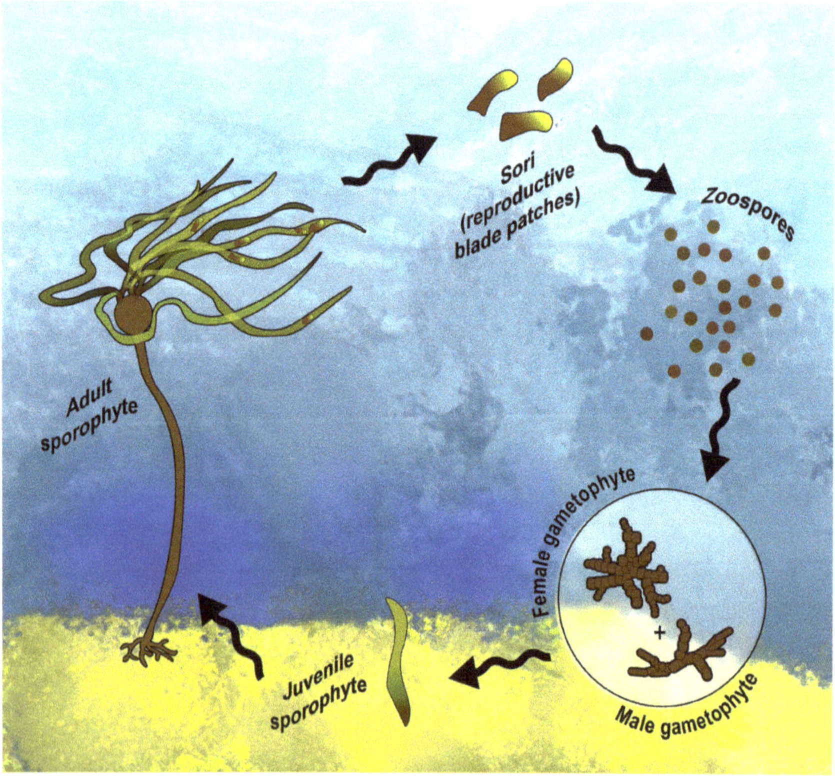

Figure 1. Life cycle of the kelp Nereocystis luetkeana. Adapted by Max and Becs from a lecture by Dr. Brooke Weigel

substrates and mature into 'gametophytes'. Gametophytes are haploid (containing only one copy of their parental genetic material), much like sperm and egg cells. When 'male' gametophytes release sperm cells, those sperm can fuse with egg cells in 'female' gametophytes to produce a diploid (containing two copies of each chromosome) sporophyte. That sporophyte eventually grows into a mature adult that can reproduce sexually.

Max's lab had previously collected reproductive adult kelp tissue, which released spores in the lab in Chicago. However, the lab's first attempts to cultivate mature kelp never seemed to work: the spores never appeared to mature, and the lab did not see any growth of gametophytes or sporophytes for a long time. The experiment in this book was therefore designed in part to answer the question, "Does the addition of clay substrates help kelp spores transition to sporophytes more readily in the lab?"

The lab also needed a way to transfer kelp sporophytes once they germinate. Live, growing kelp are difficult to get in Chicago. The process of growing spores begins in very small volumes of liquid, which increases the concentration of spores, and therefore the probability of successful mating and germination. However, such small volumes limit the growth of sporophytes as they mature. For any sporophytes that we managed to grow, we wanted to be able to better imitate the mechanical forces and nutrient conditions of the open ocean. To make sure they could transfer any sporophytes that settled onto the ceramic substrates, we pinched one side of each substrate to create a small handle. This feature was added to help answer the question, "Do kelp sporophytes grow to later life stages when transferred to larger volumes of media, under more complex conditions?"

We varied the firing temperature, silica percentage, and surface texture of the ceramic substrates (see Figure 2). The base clay recipe is an amended version of Salvador

Sporophyte: a diploid (i.e. possessing two copies of each chromosome) life stage in kelp - the sexually reproductive stage that can begin after successful mating of male and female 'gametophytes' (which possess only one copy of each chromosome and are 'haploid')

Jimenez's handbuilding terra cotta. The choice to vary firing temperature, silica percentage, and surface texture in the clay substrates was an attempt to address possible difficulties that the kelp experienced in attaching to a solid surface, mating as gametophytes, and growing as sporophytes. The firing temperature of clay affects how hard and porous it will be as ceramic. By providing substrates of different hardness and porosity, we hope that some of the kelp will be able to successfully penetrate the surface and attach securely. The chemical composition of a kelp's substrate could also contribute to or prevent its growth - because seaweeds take up silica as a nutrient, we hoped that different concentrations of silica in different substrates would allow some kelp to more successfully grow to adulthood. Silica is a common material in clay mixes so varying the quantity is in line with ceramic

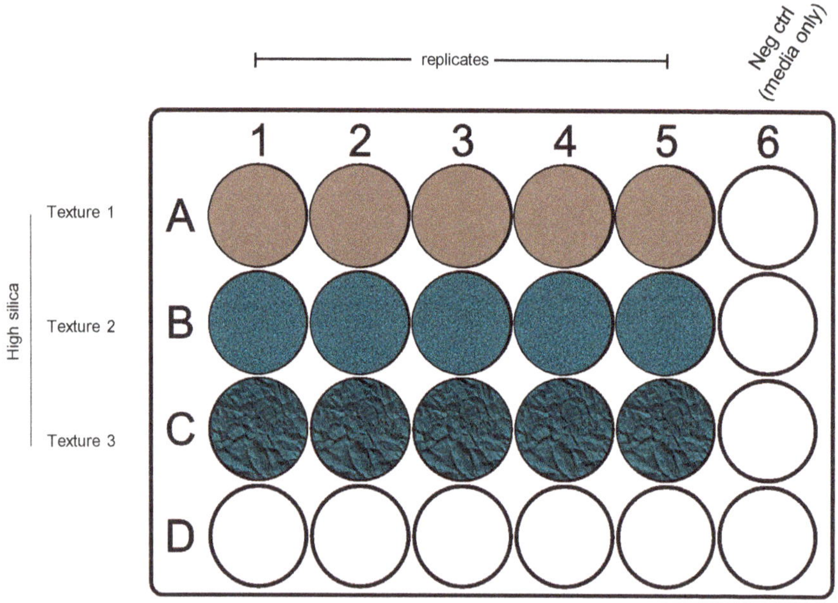

Figure 2. Experimental design for the growth substrates. Each substrate was sized to fit into the wells of a 24-well plate. Each plate contained all substrates fired at the same temperature, with the same silica content, for a total of 9 plates (3 firing temperatures x 3 silica content levels). Each plate contained 5 replicates of each texture level. The substrates were placed in the wells by hand, and a standard volume of male and female kelp gametophytes suspended in sterile-filtered seawater was pipetted into each well. Kelp were allowed to grow for two months on the plates.

understandings. Finally, the texture of the substrate could affect how easily a gametophyte settles before carrying out other life processes. We hope that the different surface textures of our substrate will offer some kelp ideal hills and valleys in which to settle and grow.

Adapted Salvador Red Earthenware Handbuilding Cone 04-1

	100lbs	200lbs
Redart	50	100
Hawthorn 20 mesh	25	50
Talc	15	30
Ball Clay OM4	10	20
Fine Grog	10	20
Medium Grog	10	20
Water	25lb	50lbs

Table 1. Base recipe for the clay body used to make the substrates.

Each step of making these substrates feels like an attempt at apology. We are trying to amend future harm by finding precisely what it means to make kelp more comfortable in a lab. This process required individually mixing three varieties of clay, taking texture samples from the environment, rolling out slabs of clay, hand cutting circles that will fit into small wells in the trays that will become the kelp's home, texturing those circles, pinching the edge to make a way to remove them from their eventual trays, and running three separate kiln firings.

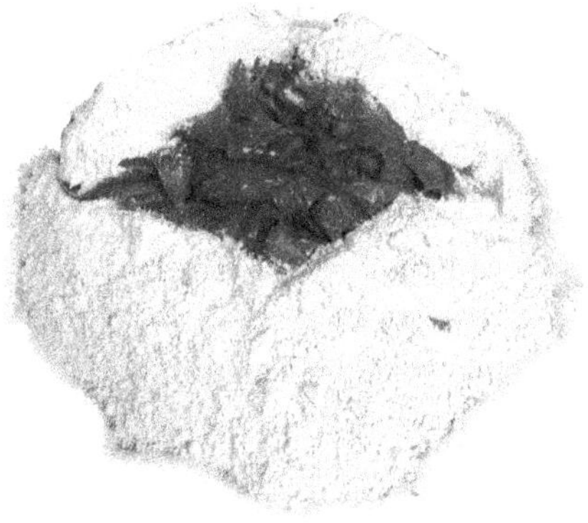

Image 26. Photo taken just before mixing the 'very high silica' clay body. Base clay recipe poured over a bed of silica powder

	Firing temperature	Texture	Chemical composition
Level 1	High temp cone 1	Tree bark	1 - very high silica - total percentage = 65%. Beginning with 4.8 lb of basic recipe, add 3.09 lb pure silica
Level 2	Low temp cone 04	Sand/grainy	2 - high silica - total percentage = 55%. Beginning with 4.8 lb of basic recipe, add 1.33 lb pure silica
Level 3	Super low temp cone 010	Smooth, glass/metal	3 - no silica added - total percentage = 42.5% DW

Table 2. Table of variables, with explanations of the levels of variation. The different firing temperatures were achieved by changing settings on the kiln. The different textures were achieved by pressing different materials into the surface of the clay before firing. The different silica concentrations (weight/weight) were achieved by adding different quantities of silica powder to the base Red Earthenware recipe. The base is 42.5% silica, and this served as our lowest silica level. To achieve 55% silica by weight, they added 1.33 lb of pure silica powder to 4.8 lb of the base. To achieve 65% silica by weight, they added 3.09 lb pure silica powder to 4.8 lb of the base. Max calculated the amounts of silica to add, and Becs mixed the powder into the base by hand.

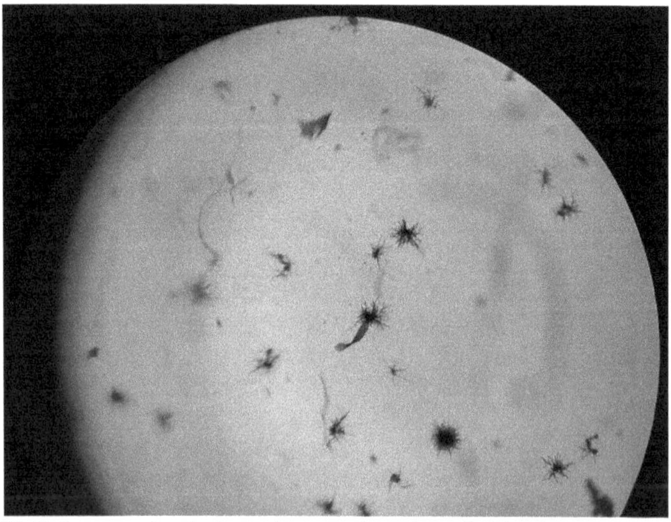

Image 27. Kelp gametophytes, observed under a light microscope

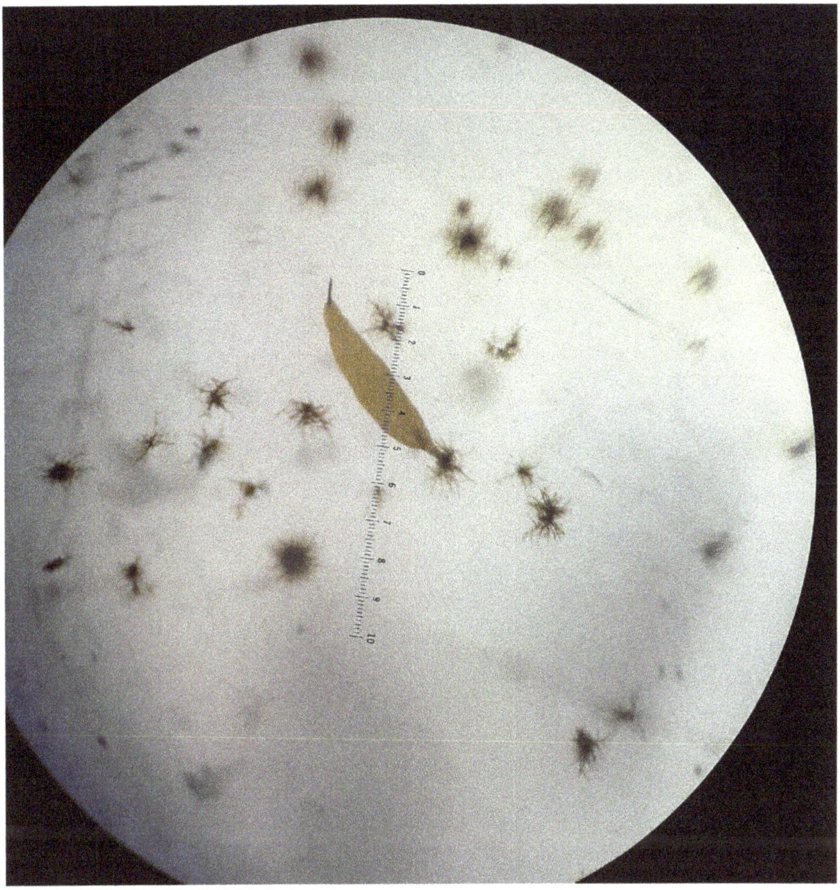

Image 28. Kelp sporophytes, observed under a light microscope and identified by the emergence of a blade

In total, we created 135 substrates (3 temperature levels x 3 silica levels x 3 surface textures x 5 replicates). The substrates were placed into nine trays. Each tray corresponded to a different combination of silica content and firing temperature of the substrates, and each tray contained 5 replicate substrates of smooth, intermediate, and rough textures. Each tray included three 'control' wells that contained no substrate. (Figure 2).

	Super low temperature (SLT)	Low temperature (LT)	High temperature (HT)
No silica (NS)	NS-SLT	NS-LT	NS-HT
Medium silica (MS)	MS-SLT	MS-LT	MS-HT
High silica (HS)	HS-SLT	HS-LT	HS-HT

Table 3. Abbreviations used in naming plates

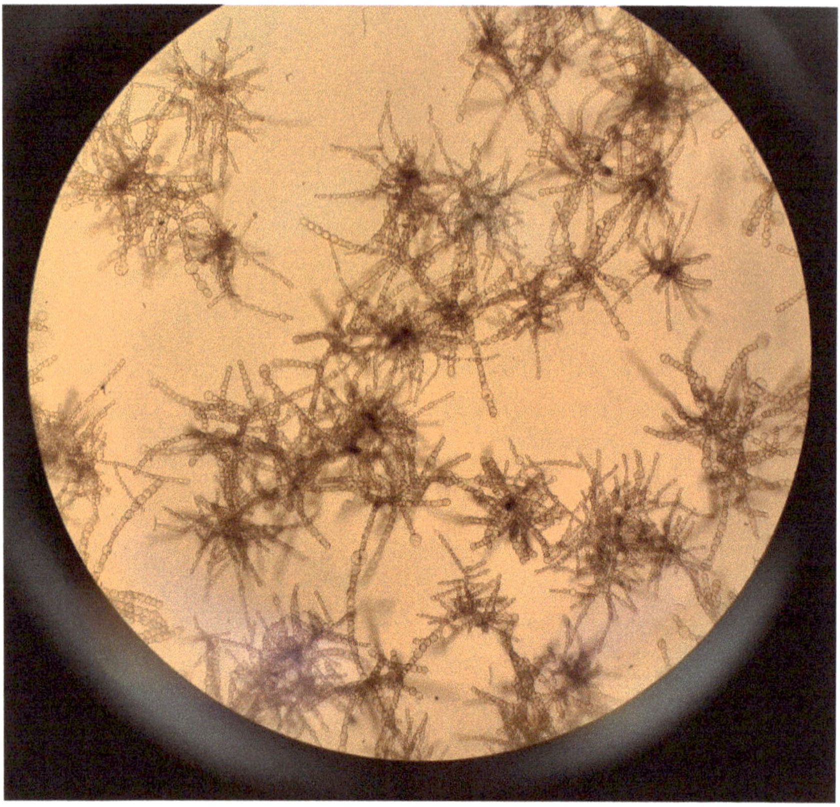

Image 29. Kelp sporophyte, observed under a light microscope

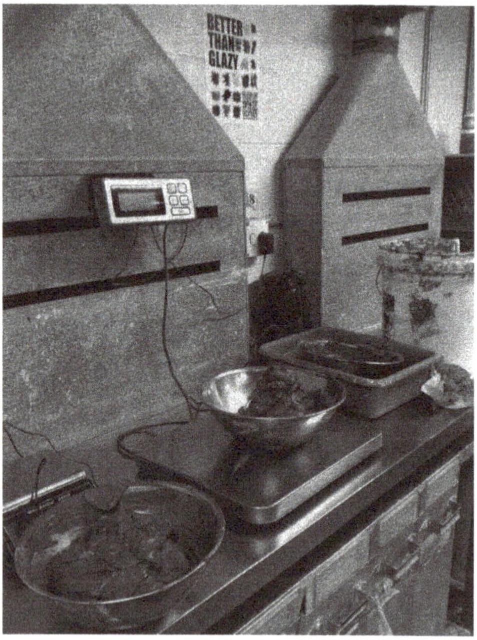

Image 30. All three clay bodies prepared in slurry form

Image 31. Photo taken while we pinched the clay substrates to create a 'handle' for lifting and moving

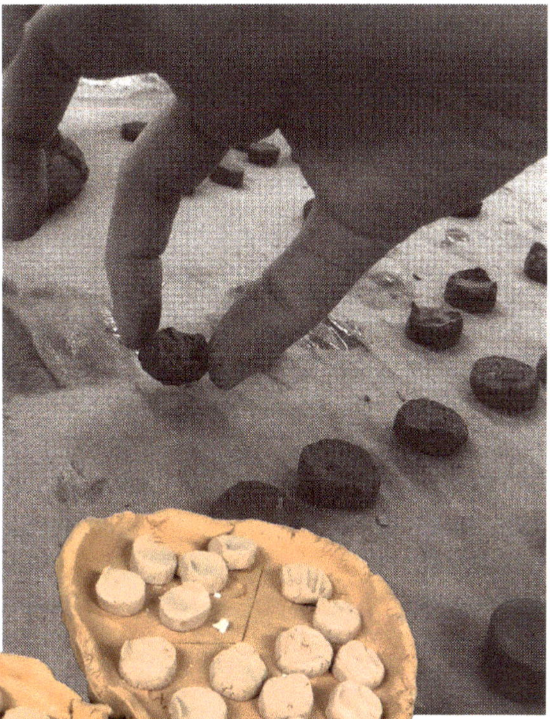

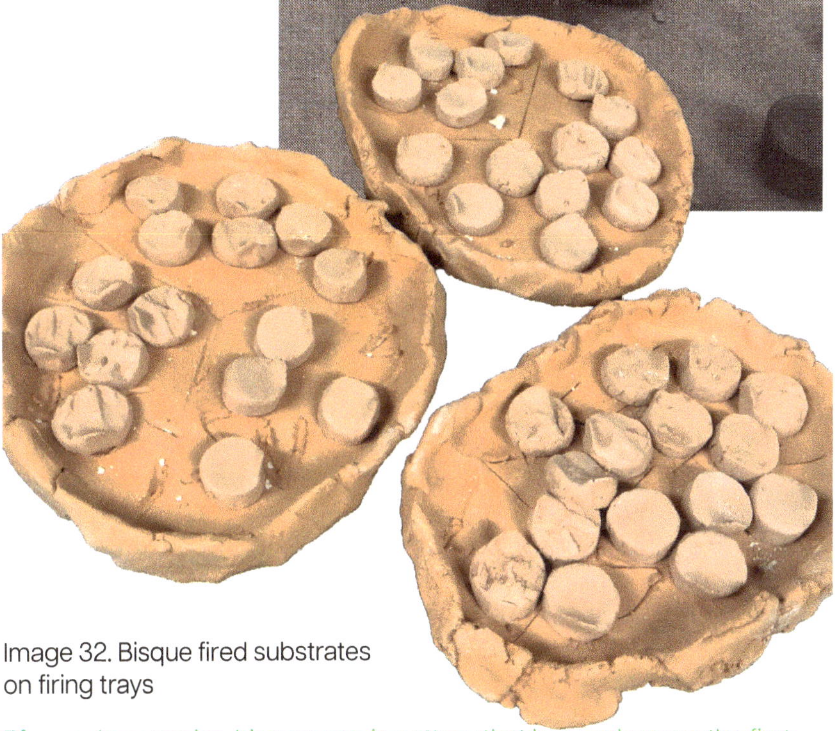

Image 32. Bisque fired substrates on firing trays

Bisque: In ceramics, bisqueware is pottery that has undergone the first firing to bisque temperatures. Bisque pieces are low-fired so that all water leaves the clay but the ceramic is still porous. Most functional ceramics gets fired a second time at a higher temperature to vitrify it and bring additional glazes up to flux temperatures.
Vitrify: vitrification means to turn into glass. In ceramics, it also means when the clay body is mature which often includes becoming denser and less porous.

Beginning the experiment and preliminary results

The substrates were placed into wells and sterilized by exposure to UV light, and then gametophytes and fresh seawater were added. Max pipetted male and female gametophytes into each experimental well beginning on April 9th, 2023.

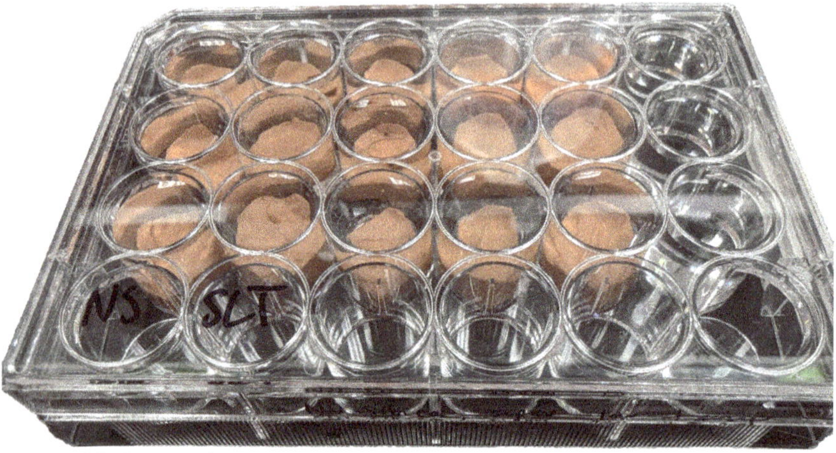

Image 33. Photo of a plate, just before sterilization

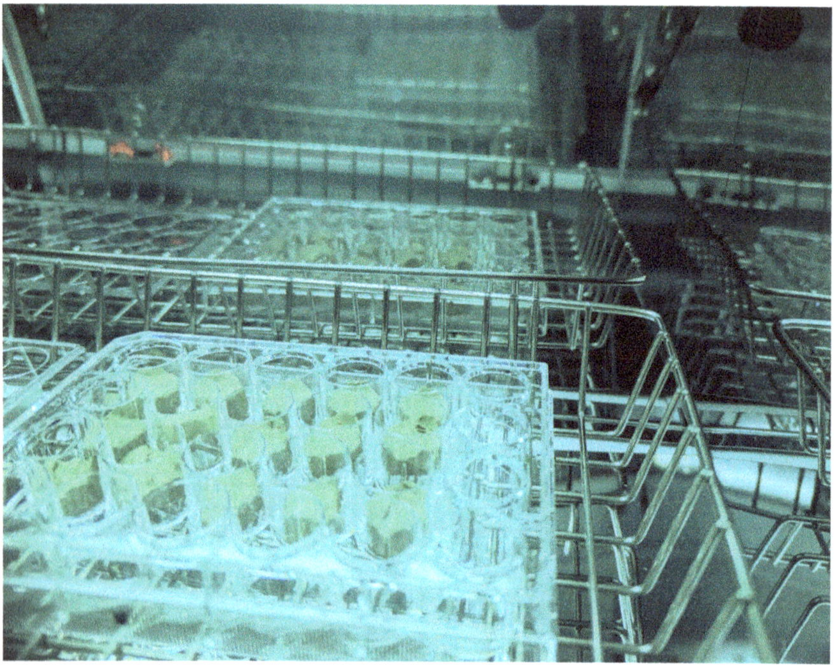
Image 34. A plate loaded with substrates being sterilized by UV light

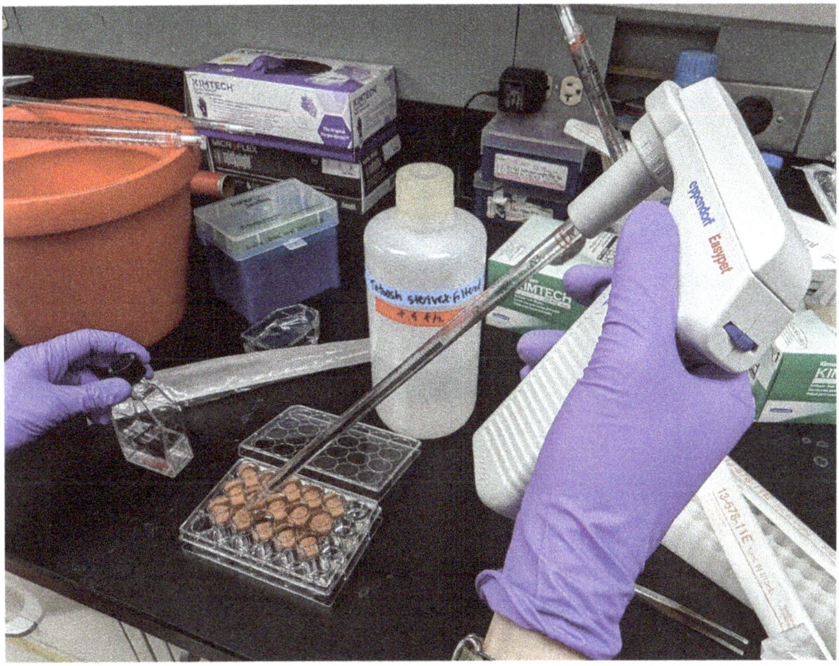

Image 35. Max pipetting gametophytes over a plate

Mid-point check-in

It is now a bit more than a month later. Each tray has at least some kelp growth, though much of the growth looks like asexual gametophytes and the kelp seem to grow slower on our substrates than on the smooth plastic of the empty control wells. Their morphology also looks different, with many of the kelp in wells with a substrate appearing more filamentous than the kelp in control wells. It is too early to draw strong conclusions about the effects of each substrate treatment on sporophyte growth - only one tray (NS-SLT, No Silica-Super Low Temperature) has visible sporophytes growing, and these still appear very small.

The lower silica content substrates have the greatest amount of growth; the high firing temperature substrates have lower growth compared to other firing temperatures.

Filamentous: Resembling a thread or fiber; containing or composed of threads/fibers. **Treatment:** Experimental condition; in contrast to 'control.'

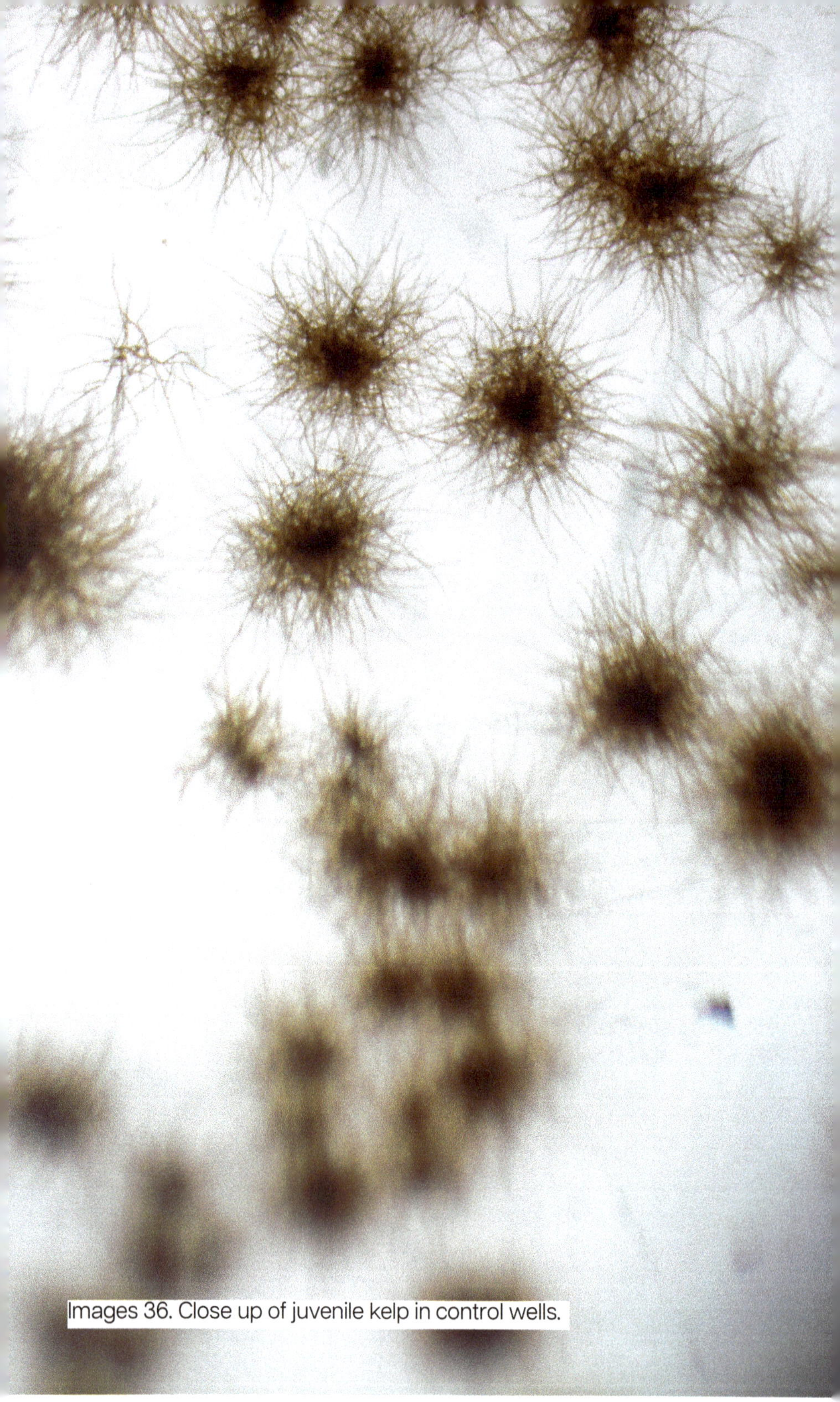

Images 36. Close up of juvenile kelp in control wells.

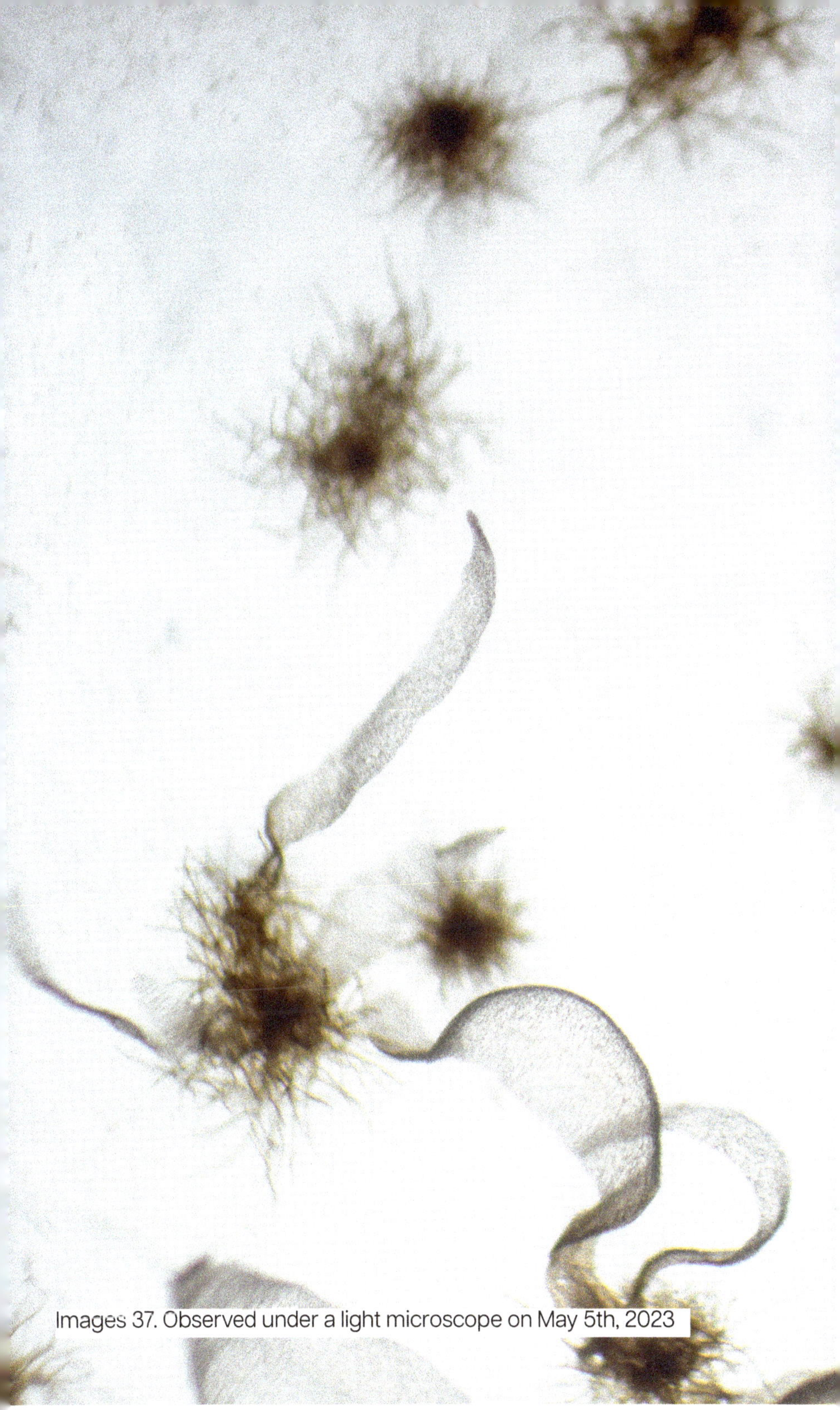

Images 37. Observed under a light microscope on May 5th, 2023

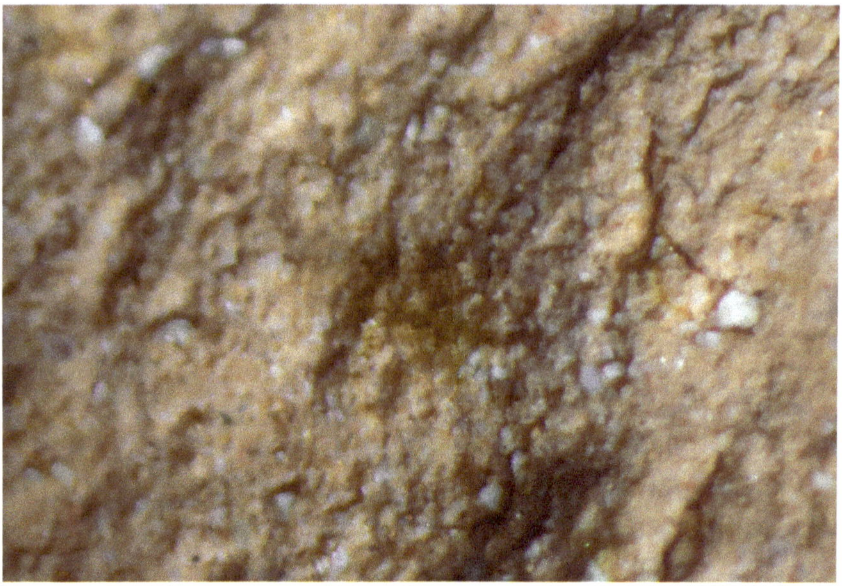

Images 38. Close up of juvenile kelp attached to a substrate

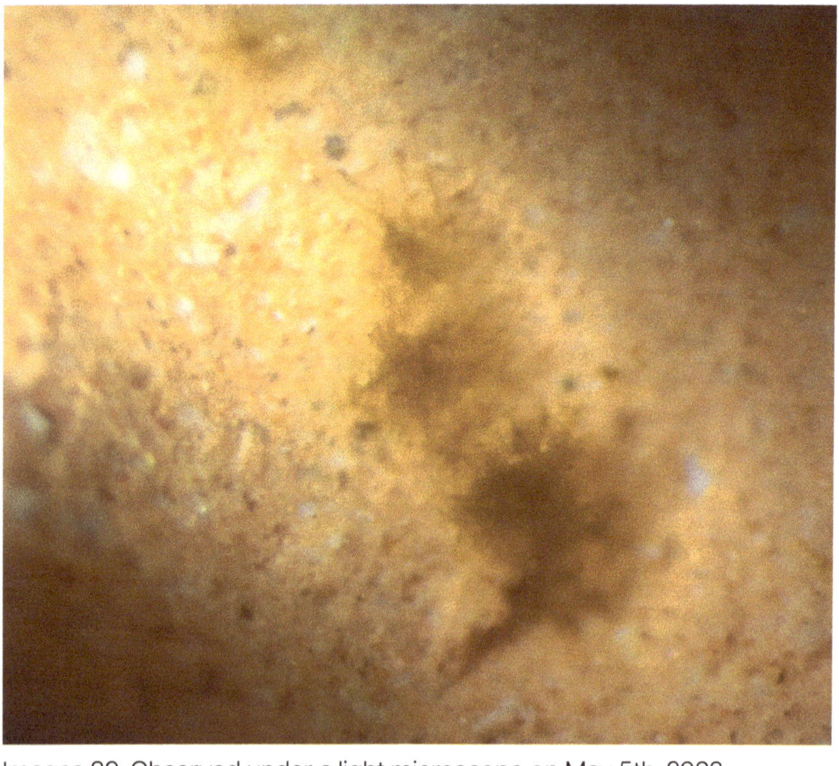

Images 39. Observed under a light microscope on May 5th, 2023

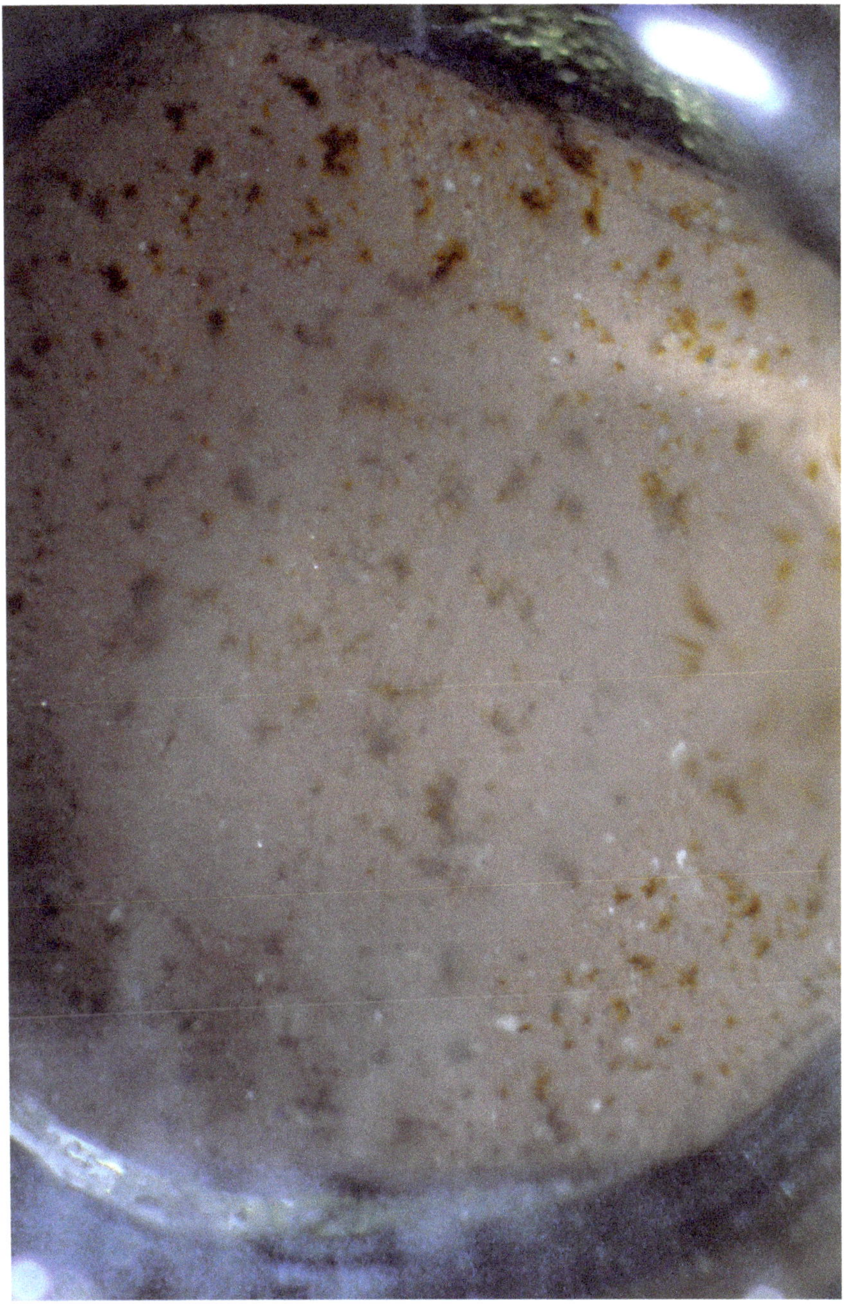

Images 40. Most of a ‘smooth’ substrate surface, showing juvenile kelp attached. Observed under a light microscope on May 8th, 2023

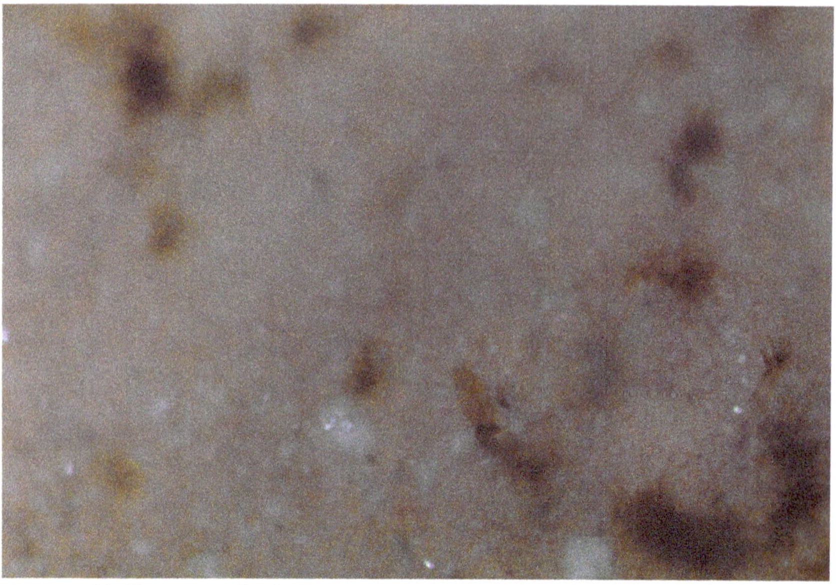

Images 41. Close up of a kelp sporophyte

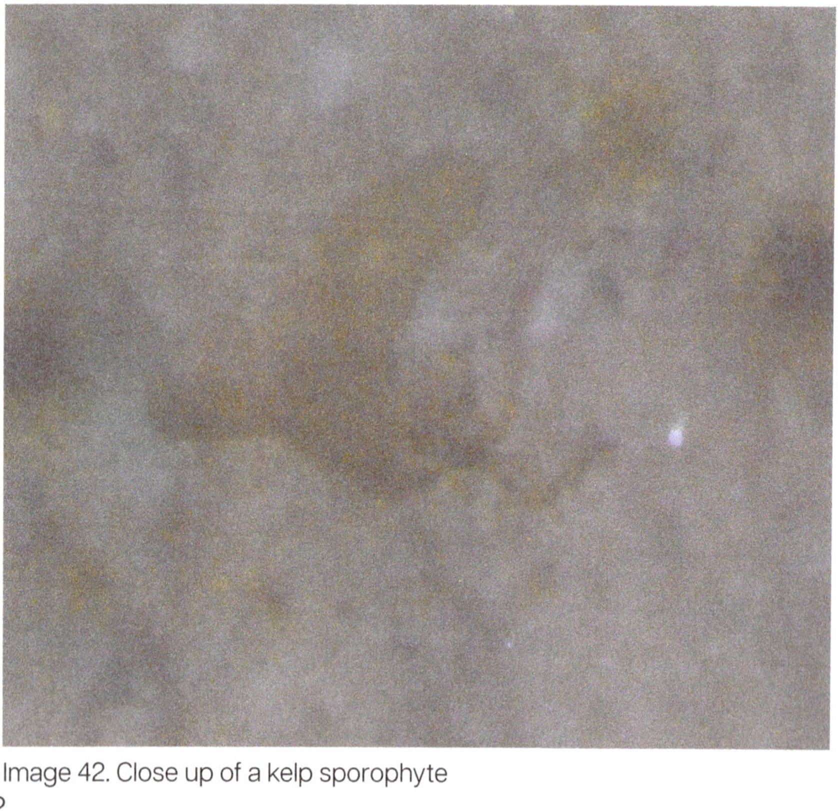

Image 42. Close up of a kelp sporophyte

Our next steps are to keep an eye on the substrates as the kelp develops. Once we find the optimal substrate material, Becs will make sculptural kelp-size furniture with those variables.

Statistical analysis

The experimental design included three variables of interest, each of which represented a hypothesis about the most productive way to apologize to kelp. To measure the effect of each variable on the kelp gametophytes, we counted the number of sporophytes with visible blades on all substrates, and in 'control' wells that lacked clay substrates. We used this proxy for maturation because blades form only after the transition to macroscopic sporophyte.

We divided the number of sporophytes counted on each substrate by the number of sporophytes counted in the 'control' well (which contained no substrate) in the same row on the plate. This procedure was intended to account for growth effects that might be due to spatial arrangement of the substrates on the plate. We also alternated who counted the blades (Becs or Max) by row, to try to randomize our observational biases.

The normalized abundance of sporophytes was calculated for each replicate (5 replicates per combination of silica content, firing temperature, and surface texture), and plotted using base R. Significance was determined by applying the Kruskal-Wallis Rank Sum test in base R.

Results

Silica content

We hypothesized that higher silica content in the substrates might promote sporophyte germination. Kelp (and other photosynthetic organisms like plants) accumulate silica, and this metabolism is thought to be linked to carbon sequestration. Though the biochemical mechanisms of silica uptake are poorly described in the literature, we reasoned that increasing silica content might provide a limiting nutrient to the baby kelps in our lab. Counter to our expectation, we found no significant relationship between the silica content of the substrates and the number of sporophytes counted. In fact, the trend was that lower-silica substrates tended to host more sporophytes counter to our expectation. The uninterpretable error bar on the 'LS' treatment is due to our

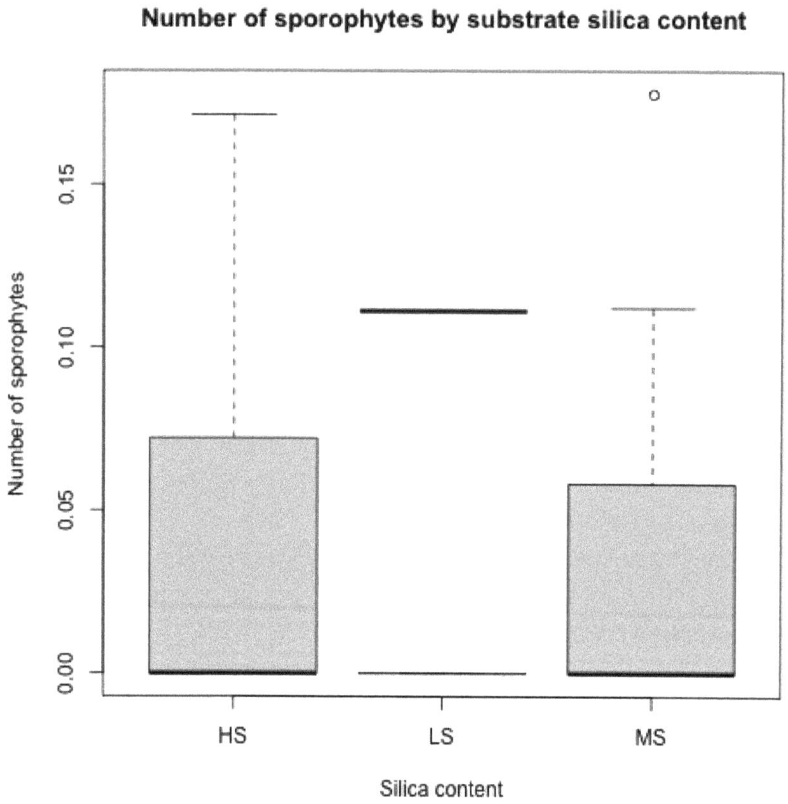

Figure 3. Normalized count of sporophytes observed on substrates, separated by substrate silica content.

normalization - for some 'LS' substrates, we counted zero sporophytes growing in the corresponding control wells. This leads to a 0 in the denominator of our 'abundance' for some samples, and apparently no variation (no error in measurement) for those wells.

Substrate texture

We also hypothesized that the texture of the substrate surface might influence the ability of the gametophytes to settle, mate, and germinate. Rougher textures create greater substrate surface area, possibly allowing for a greater density of gametophytes (both male and female) to settle at close

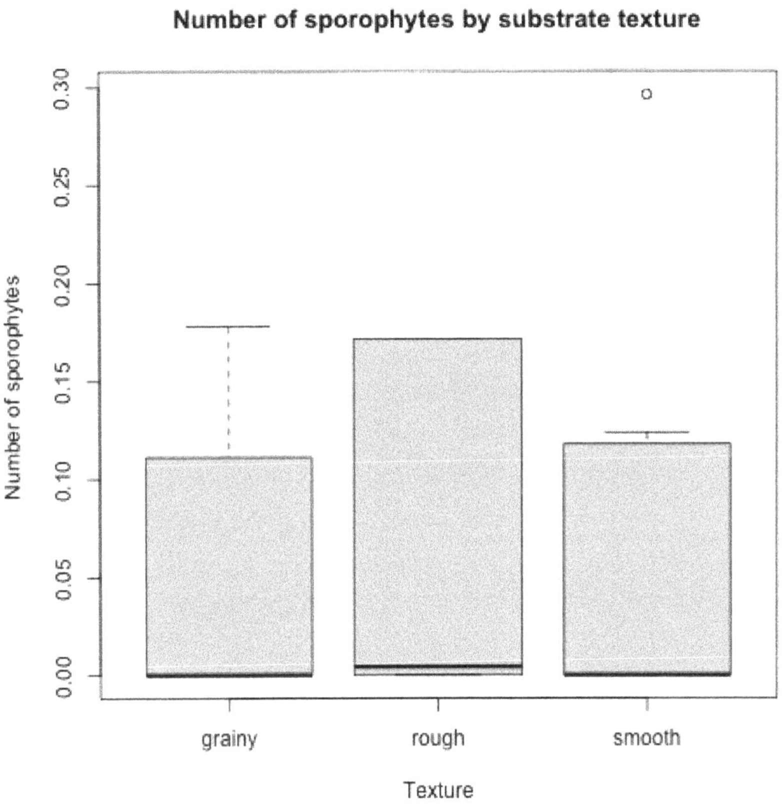

Figure 4. Normalized count of sporophytes observed on substrates, separated by substrate surface texture.

proximity. Again, our results contradicted our expectation: while there was no significant relationship between substrate surface texture and the number of sporophytes counted, the smoother textures tended to host more sporophytes than the rough-textured substrate.

Firing temperature

Finally, we hypothesized that the temperature at which the substrates were fired might influence the likelihood of sporophytes attaching to the substrates. Firing temperature affects the porosity of the ceramic, and perhaps a more porous substrate has more surface area and more microscopic 'valleys' into which gametophytes could stably settle. We observed significantly more sporophytes on substrates fired at super low temperature, relative to either low or high temperatures.

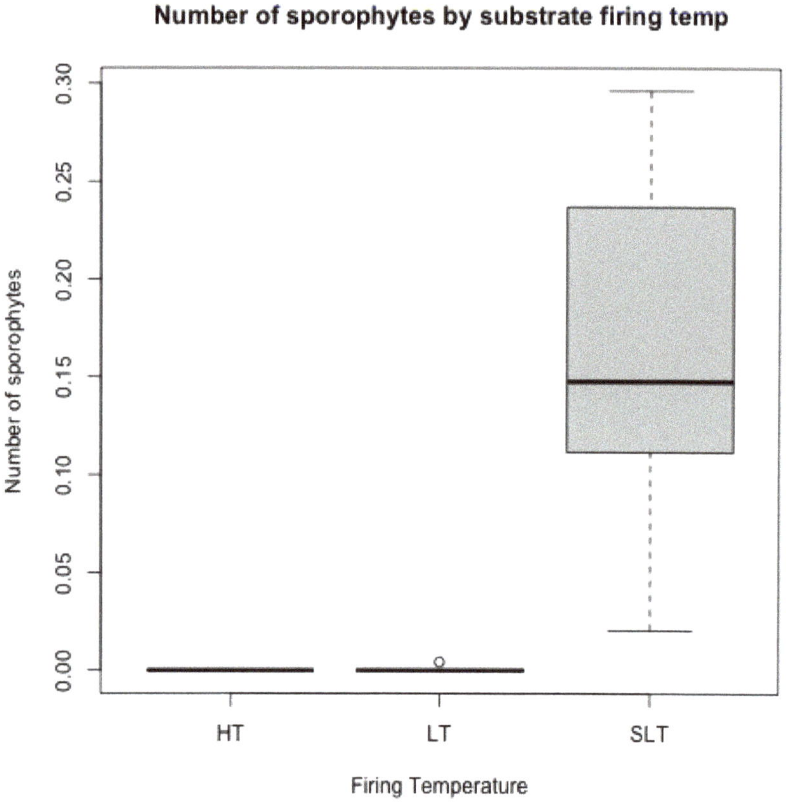

Figure 5. Normalized count of sporophytes observed on substrates, separated by substrate firing temperature.

Cone: Pyrometric cones measure the temperature ceramics reach in a kiln. They are physical cone shaped objects that bend and eventually melt as they are heated to specific temperature range. Ceramic firing temperatures are referred to by cones starting from low at cone 022 to the hottest at cone 14. This heat work measurement is a combination of heat and time so there is not an exact temperature for each cone.

A

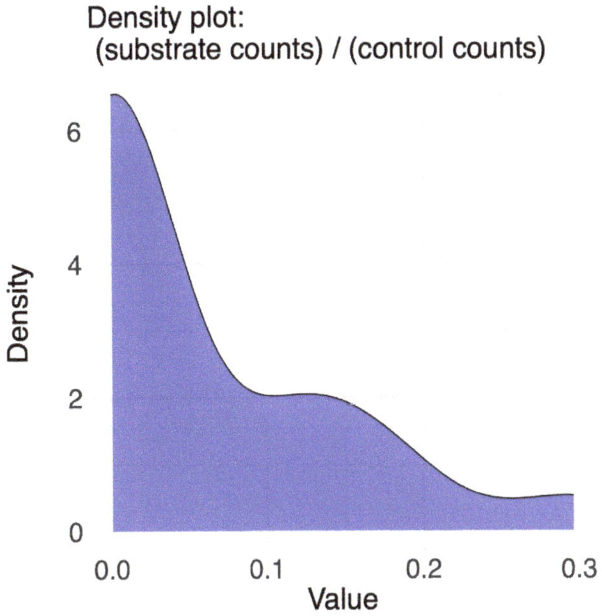

B

```
> ######kruskal wallis########
> kruskal.test(avg ~ SilicaContent, data = data)

        Kruskal-Wallis rank sum test

data:  avg by SilicaContent
Kruskal-Wallis chi-squared = 1.4643, df = 2, p-value = 0.4809

> kruskal.test(avg ~ Texture, data = data)

        Kruskal-Wallis rank sum test

data:  avg by Texture
Kruskal-Wallis chi-squared = 0.33652, df = 2, p-value = 0.8451

> kruskal.test(avg ~ FiringTemp, data = data)

        Kruskal-Wallis rank sum test

data:  avg by FiringTemp
Kruskal-Wallis chi-squared = 12.364, df = 2, p-value = 0.002066

> kruskal.test(avg ~ Counter, data = data)

        Kruskal-Wallis rank sum test

data:  avg by Counter
Kruskal-Wallis chi-squared = 0.24914, df = 1, p-value = 0.6177

> kruskal.test(ctrl ~ Counter, data = data)

        Kruskal-Wallis rank sum test

data:  ctrl by Counter
Kruskal-Wallis chi-squared = 0.016887, df = 1, p-value = 0.8966
```

Figure 6. Distribution of normalized substrate counts across all treatments. The distribution of our data was non-normal, leading us to choose the Kruskal-Wallis Rank Sum Test to test for significance of effect. A. Histogram of counts. B. Results of Kruskal-Wallis test.

Interpretations

Our experiment indicates that ceramic substrates fired at super-low temperatures of cone 010 (around 1641°F) facilitate higher rates of sporophyte germination than similar substrates fired at low (cone 04 or around 1940°F) or high (cone 1 or around 2109°F) temperatures. We hypothesize that this is due to the higher porosity of SLT (Super Low Temperature) substrates compared to the other ceramic surfaces. We imagine that, for invisibly tiny gametophytes, pores even a few micrometers in diameter might create appealing, sheltered spots to settle. Perhaps this would lead to denser colonization of gametophytes onto the SLT substrates, and a higher probability of successful mating.

After our final sporophyte count on June 2nd, 2023, we moved the substrates that hosted sporophytes into larger volumes of liquid. We also added steady oxygen flow and regular shaking to perturb water flow. When we attempted to move the "control" sporophytes (attached to the plastic plates) into similar new conditions, the sporophytes did not reattach and they quickly died. We report anecdotally that the substrate-attached sporophytes that were transferred to more complex conditions developed larger blades and appeared to have differentiated tissue structures compared to sporophytes that remained attached to the plastic of the control wells. However, because the control-well sporophytes remained in small volumes of liquid with no oxygen flow, we cannot attribute any observed differences to the substrates alone.

Limitations of the study

An important step in rigorous research is considering the limitations of your method. As gestures of apology, the ceramic substrates achieved our goals. We identified a feature of the substrates that appears significantly associated with different rates of sporophyte germination. The substrates also allowed us to transfer sporophytes into new laboratory conditions that more closely resemble the complexity of their natural environment. These substrate-attached kelp grew to a larger size than any other sporophyte grown in Max's lab before or since.

As an intervention in kelp husbandry, however, the substrates will require more testing before we can recommend that they be incorporated into kelp husbandry. Labs grow kelp for a multitude of reasons such as repopulating certain decimated areas, learning what kelp needs to survive better in its natural environment, and to understand what the short term future of kelp looks like. Max's lab needs to grow kelp to ask questions about how kelp-associated bacteria affect early-stage kelp development. Max and Becs take these scientific goals seriously, and bear them in mind when discussing limitations below.

Limitation 1: low reproducibility

In part due to time and financial resources, our study involved a small number of substrates and sporophytes counted. Even though we detected a significant effect of substrate firing temperature on the rate of sporophyte germination, the observed outcomes represents a limited subset of all possible outcomes. For example, if we conducted exactly the same experiment, but under different temperature conditions, perhaps the observed effect would hold and perhaps it would not. Our results have not yet been shown to be reproducible, which makes it impossible to prove that our observations were not the result of the exact conditions of the laboratory in which they were conducted. While this does not detract from the value of the experiment as a gesture of apology, it limits the generalizability of our claims about kelp husbandry.

Let us imagine that, in reality, there is absolutely no effect of ceramic substrates on kelp growth. Because the world is large and mysterious, it's possible that the number of sporophytes we observed and the pattern of their distribution was due to chance alone, or bias in our method. We used p-values and the Kruskall-Wallis Rank Sum Test to quantify the probability that we would observe the relationship between our variables (silica content, firing temperature, surface texture), even if no 'true' relationship exists. In this test, we compare the number of sporophytes counted on substrates of different conditions and ask: how different is the distribution that we observed from a 'null' distribution, which we might expect if the substrates had no effect on sporophyte growth? In our case, a 'null' distribution would be one where there is no relationship between the substrate added and the number of sporophytes counted - in other words, a truly random distribution of sporophytes across both substrates and controls.. The Kruskal-Wallis test showed a significant effect of substrate firing temperature, with a p-value of 0.002066. This p-value suggests that, if the 'null' expectation of no relationship between substrates and kelp growth is really true, then the distribution of sporophyte growth that we observed would only occur 0.2% of the time.

This statistical test answered a very specific question: how did the addition of a complex growth substrate affect juvenile kelp development, in the Pfister lab at the University of Chicago during the Spring of 2023? However, we tested only 3 'growth-related' variables in total, and counted only 532 sporophytes. This is a vanishingly small fraction of all possible sporophytes, and all possible variables that could affect sporophyte growth. In the future, we believe that clay substrates should be tested further as substrates for laboratory kelp growth. Repeating the experiment with a larger number of substrates and independent experimental conditions would allow us to confirm our findings related to firing temperature, and possibly detect other, subtler connections between juvenile kelp and their laboratory growth environments.

Limitation 2: operator bias

The ultimate measurement in our experiment was the number of sporophytes on each substrate, and we took this measurement by placing all substrates under a light microscope and counting by eye. This method is subject to 'operator bias' in that Becs and Max have inherently different eyeballs and propensities to identify something as a sporophyte - though we examined sporophytes together so that we would share an understanding of what sporophytes look like, we could never be sure that our criteria for what 'counts' was exactly the same. We alternated 'counter' by row and tried to randomize according to silica content and firing temperature of the substrates in order to minimize the effects of this bias.

To quantify the difference in our average tendencies to see and count sporophytes, we plotted the number of sporophytes counted by Becs and Max on substrates and 'control' wells with no substrates. Max on average counted more sporophytes on substrates, though the difference was not significant (p-value = 0.6177). Max counted fewer sporophytes in control wells than Becs on average, though the difference was not significant (p-value = .8966).

Limitation 3: statistics can only tell us so much in the context of apology

Most research methods in the natural sciences undergo extensive statistical validation before they are implemented en masse. This is a crucial step in our practice of knowledge-building, because statistical tests help us quantify the 'fit' between the data we collect, and the data we might expect to observe based on our broader knowledge of the

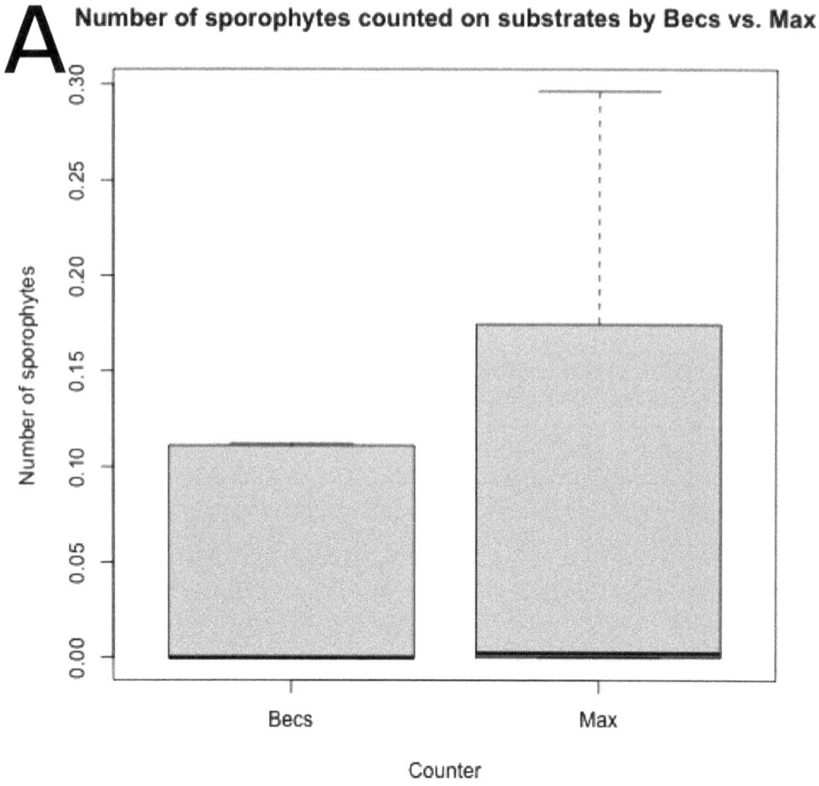

Figure 7. Distribution of normalized substrate counts, separated by counter. A. Distribution of sporophytes counted on substrates, by counter. B. Distribution of sporophytes counted in control wells, by counter.

world. If a research method consistently results in unexpected observations, a 'good' natural scientist will question the validity of the method, or the validity of the expectation, or both.

In our study, we manipulated the growth environment of juvenile kelp, with the expectation that certain variables would affect the germination and longevity of these young specimens. We applied statistical tests to quantify the effect of three variables in particular and found that substrates fired at super low temperatures facilitated sporophyte germination relative to other substrate conditions. These tests are powerful in that they validate our hypotheses about what kelp needs to grow: they help us model the shape of the world and the connections between kelp and environment. But they are also limited: it can be difficult for both kelp and

B

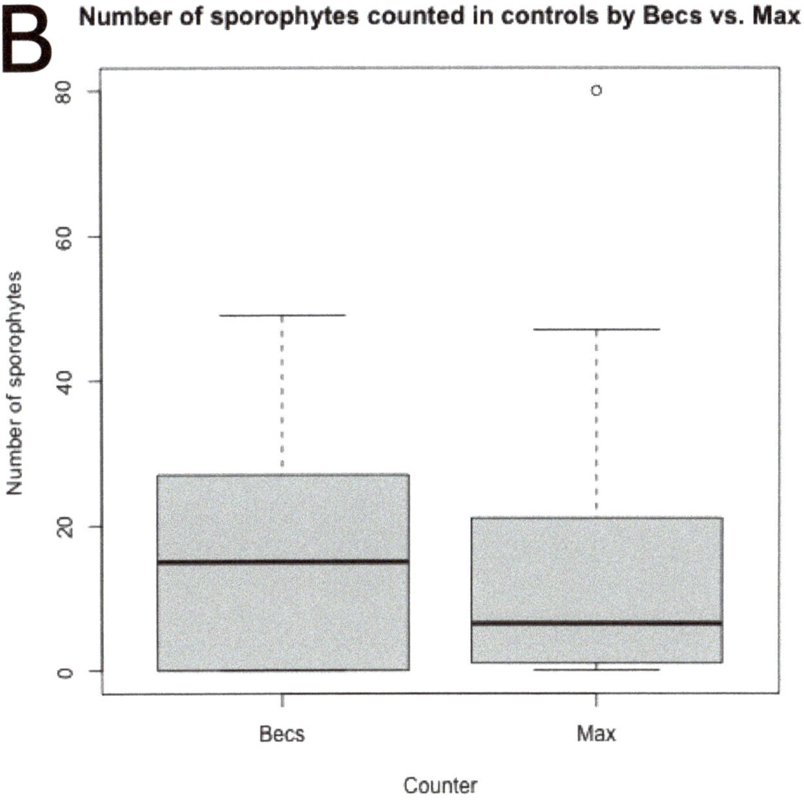

humans to communicate nuance in the language of statistics. We interpreted our statistical analyses with the understanding that they will always be imperfect mediators of bi-directional human-non-human communication.

We joined our disciplines with this experiment in an attempt to make kelp more comfortable in a lab setting. This somewhat wholesome and domestic phrasing translated into creating a human-made environment that enabled the kelp specimens to "grow better." In this task, we succeeded. We were able to figure out a way through a bottleneck in kelp growth at Max's lab. We were trying to find a solution to this issue: kelp kept growing on the plastic petri dish instead of the rocks that were meant for them.

SO what's the problem with that? There was no way to then transfer the kelp to a larger container with oxygen flow to allow growth to maturity. Kelp would form on the petri dish and die there. We created ceramic substrates that could be easily picked up and moved to the larger, more dynamic environment after starting growth in the sheltered container. Bull kelp forms a holdfast which is an attachment often to a rock or other growth in the sea. In this lab setting, we helped the kelp form permanent holdfasts to the ceramic substrates. Our downfall was a lack of money and support to test these substrates in larger numbers and with more attention to the kelp's survival after the move to a bigger tank. If a team with optimal funding and support attempted to replicate and push forward this idea, ceramic substrates could be a major innovation in kelp growth.

This entire book serves as an attempt at apology to kelp from a scientific and artistic perspective. This physical experiment is our most embodied form of apology that we have attempted. We combined our art making experience with our understanding of kelp and lab processes to create a new way of being with kelp. Through this experiment, we met kelp's needs at a higher level than before. Even though eventually our specific specimens shriveled up and died (which required another apology), this was a successful scientific experiment and a successful artistic piece and a successful apology.

Apologizing to Kelp

Becs on Apology to Kelp

Collective apology is powerful and potentially transformational. As we invite you into an *Apology to Kelp*, as we ask you to see the ocean as home, I want to also invite you into my dreamscape. This book comes out of my year long practice of earnestly apologizing to a plant every, single, day. That artistic ritual began with a dream. I drifted into a mid-day sleep and I saw a brilliant, bright yellow light above me. As my hands reached towards the light, the yellow flowed into my fingers and then my arms and with it came the most wonderful warmth, like having a home cooked meal and sitting the exactly right distance from a fire. With that warmth, followed the most incredible happiness, like floating, like flying, like being untouchable. That yellow, warm, joyous light told me to apologize specifically, and sincerely to a plant everyday for a year. I began on March 3rd, 2022. I confide this vision onto the printed page in hopes that you, my hopefully kin, can read between Max and my words, our experiments, and our microscope pictures. Go along with us on this marvelous adventure into being kin with kelp. Being kin means not just accepting but embracing the non-human as interlinked with us non-hierarchically. With that, the following essay is myself whispering and yelling at the same time and just trying to show that I am sorry.

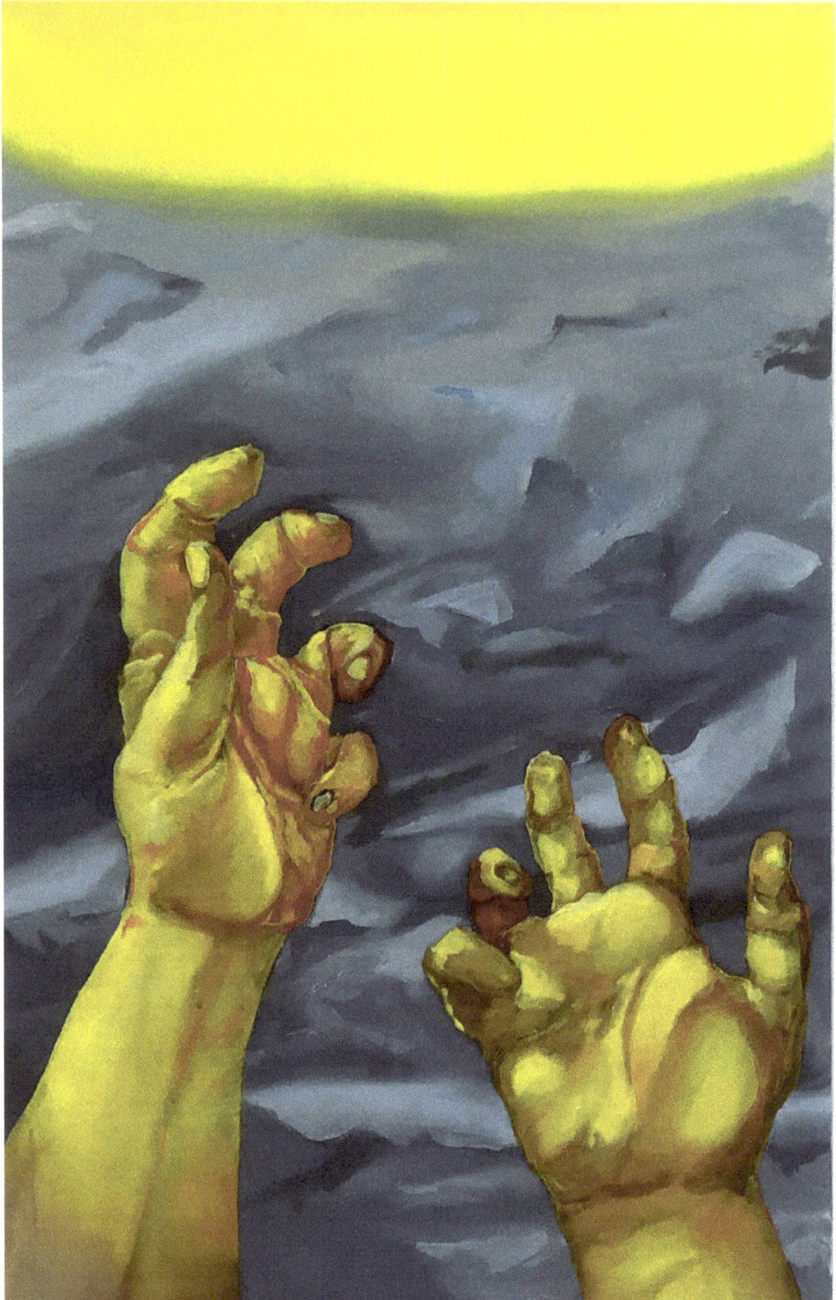

Image 43. *A Vision* by Becs

A Year of Apologies: Between the Lines

As I perpetually apologize, I see the tendrils of regret, guilt, and hope twirl between ceramicized apologies and my annotated calendars. Can you see the direction the tendrils are morphing? I am worried that no one else can see the wind that blows through my practice. So let me show you: out of my grid-like calendars, tendrils of air move, shivering blue and green; like a CGI animation reaching for the objects that they embody. As May relates to July, boxes three and 28 bridge the gap of white wall by swooshing out a stream of turquoise, from that flow breaks off a new stream to the ceramic panel six feet away. As I look at my installation, I see a blue-green dust storm straight out of Fantasia animation studios. The wind follows my past, present, and future intentions with the slow moving tentacular vines. If you still can't see what I see, flip to any page in a book: read the written apology out loud while you trace the daily drawing with your finger, your lips, let your senses be filled with desire.

What do you desire? I desire to be thoughtful, to love the way I wish to be loved. To protect those with all my heart even when I am not brave. Squint a little, close your eyes, let the nanobots of plantal healing fly at hyperspeed through your arteries. Open your eyes. The tendrils shoot (drip) from your eyes as you become entangled with apologies. What once looked to you like a whitebox gallery installation is now laced with a Mycorrhizal network clove hitch knotted. When you breathe in, the spores coat your throat. You don't feel any better, your conscience weighs just as heavy or heavier as when you picked up this book and entered my world.

Welcome by the way, I forgot to say that, I've forgotten how to host humans. And yet, as you will see/seen, I don't know how to host plants either. The point is that I am a human and I am trying my best and that is beautiful, and that is powerful, and that I need to do this. I had a vision in December of 2021. All I can do is promise, I pinky swear on Donna Haraway, Wynn Alan Bruce, and the roses from my childhood garden, I promise to keep courting failure, to understand my relationship with the botanical a pinch better.

Max on Apology to Kelp

One of my favorite TV shows in elementary school was Animal Planet's The Most X-treme: a series dedicated to weirdest, wildest ways that organisms adapt to their environments. Growing up in the Oklahoma suburbs, the unique adaptations of hammerhead sharks, blue-ringed octopuses, and star-nosed moles seemed fantastical, almost alien. To me, the work of this book is an attempt to unalienate, as a necessary first step to apology.

The diversity of life on earth is all but unfathomable. Humans and sharks and moles represent a miniscule fraction of all the spectacular, mundane, creepy, and beautiful phenotypes observed in nature. In this book, we consider this diversity not as a symbol of distance between us and other species, but as a symbol of interconnectedness. From humans to octopuses to kelp, we are members of the same family tree, rooted more than a billion years in the past. In writing this book I want to remember this history, and to celebrate the web of interdependencies it implies.

I bring this up in part to explain my appreciation for the non-human as a thing that can be divorced from my cynicism about the human. Both emotional experiences compel me to apologize - but they are different flavors of apology.

In appreciation of the non-human: I've felt exuberant joy at finding an unusual mushroom, and at seeing a prairie dog's head pop out of the dry packed soil of the Oklahoma plains. With that joy comes an equal and opposite sense of grief – the mushroom and the prairie dog are here today, but who's to say they won't be next on the long list of casualties of ecosystem destabilization? To my most cynical self, it feels like they are already dead. The apology I want to make in this context is really just words of compassion, like telling a funeral-goer, 'I'm so sorry for your loss.'

Phenotype: an observable characteristic (or set of characteristics) present in a given organism. Brown hair is a phenotype; carrying venom or toxins is a phenotype; expressing a certain gene is a phenotype; the list is unending.

I have other apologies to make: apologies for the exploitation of kelp to fuel the endless march of human resource consumption, for our refusal to legislate for harm-reduction, for failing to act on principle, for ordering from Am*zon. I make these apologies because I feel guilty, or resentful, or ashamed. These apologies cannot bring transformative reconciliation: theoretically interrogating the barriers between human and non-human is simply not enough to make up for the damage that is already done and will continue to be done to kelp. These apologies and promises to try to enact change are more like confessions, to my personal shortcomings as well as to the collective sins of the systems I am a part of.

Selected Apologies

Josie Iselen *(Artist/Author/Designer/Kelp Lover/Teacher)*:
Short bio on page 22

Dear Kelp, Your majesty and beauty have allowed me the honor of getting to know some of your cohort in a small and simple way. I apologize for not letting you live at peace in the ocean without added stress. I apologize for intruding into your world with ignorance and hubris. I only hope you can forgive me for thinking you might even want to be connected with me in any way. With all the respect I can muster, Josie.

Seah Hoban *(Scientist)*:
Short bio on page 22

For living a life so often and so divorced from nature, and for my thoughtless consumption of all that is taken from nature, and so much pollution now absorbed by the oceans, like bad blood, I am deeply sorry and sorrowful. May I have the wisdom to make amends.

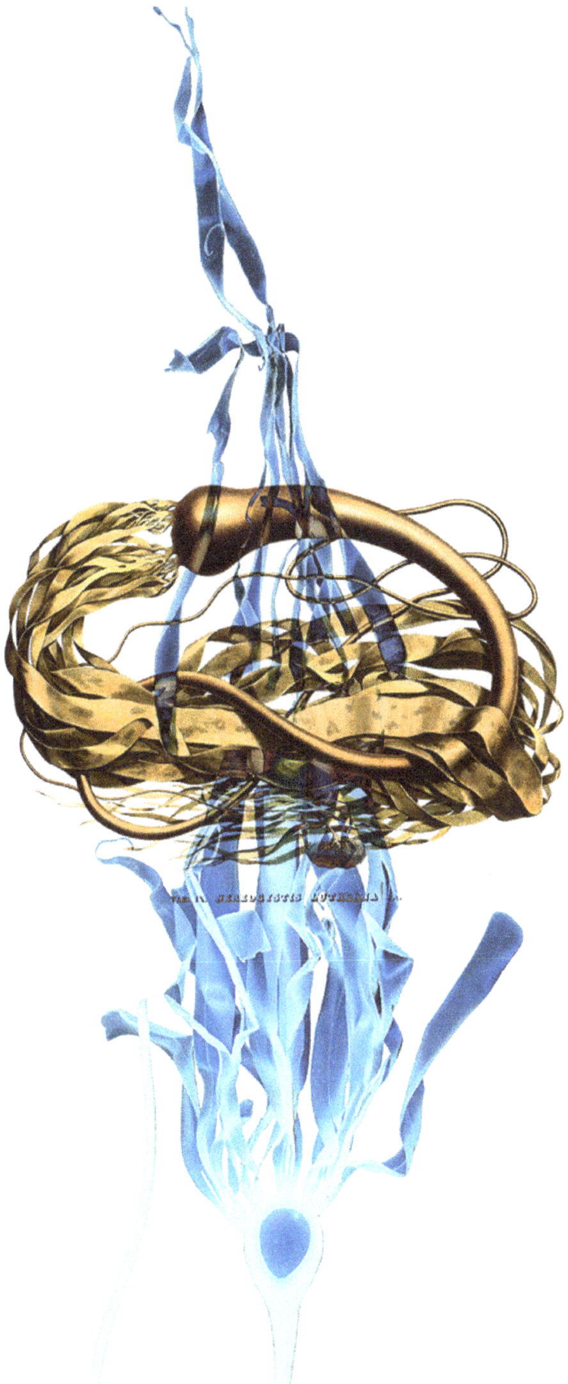

Image 44. *Nereocystis Postels* by Josie Iselen

Vince Phan *(Visual Artist)*:
Phan is a multidisciplinary artist producing short-lived artwork. His work explores the interplay between the biological and the inanimate, using both as form and material. Often working with decaying or perishable materials, Phan embraces the collateral beauty and ecology of death while simultaneously exploring ways to decolonize death.

Dear kelp,

I'm sorry.
I'm sorry that you are suffering.
I'm sorry for just learning about your sufferings.

I'm sorry.
I'm sorry for taking you for granted.
I'm sorry for not being your voice.

I'm sorry.
I'm sorry for not fighting harder for your survival.
I'm sorry for not fighting harder for our survival.

On behalf of humans, I'm sorry.
I'm sorry that the climate is getting warmer.
I'm sorry that we destroy your ecosystems.
I'm sorry that we destroy our ecosystems.

I'm afraid.
I'm afraid it is too late.
I'm afraid you will be gone soon.
I'm afraid we will be gone soon.

...

Johnny Doley *(Artist)*:
From the cicadas of his Virginian childhood, to the lettuce munching snails of his Italian garden, Doley makes art in the company of critters. An avid birder, he spends his mornings at the local marsh, dodging Red-winged blackbirds. Doley holds a MFA from School of the Art Institute of Chicago and a BFA from Boston University.

Do you remember, when
we wiggled through the thawing sea,
flagella tangled
and learning how to love?

I'll never forget when our toes touched
and I mistook yours for lungs
and ran behind my father's legs
before throwing you back to the sea.

I'm sorry I never call.
But I've been busy with tragedy–
arms and feet rooted to the land.
But still I dream to fly
through your forests
like mourning doves through the oaks
that taught me to ride a bike.

Samantha Kish-Levine *(Scientist/Student)*:
Kish-Levine is a PhD student studying evolutionary biology at the University of Chicago. Her research concerns the evolution of parasitic plants.

Dear Kelp,

I am sorry for the thoughtlessness of my species. I am sorry that our careless acts of greed warm your waters and pollute your environments.

I am sorry that I do not think of you more. You are often overlooked. You shouldn't be. Your towering stalks are majestic, your waving blades are ethereal. We do not appreciate what we have until it is too late.

I hope it is not too late.

Rochelle Voyles *(Visual Artist, Prop and Set Styling Assistant, Art Fabricator)*:
Voyles is a Brooklyn based multi-disciplinary artist whose practice oscillates between working in traditional collaged works on paper and large scale sculptures.

When I googled Kelp all of these pump-up think pieces about its possible uses came up, featuring city council members, chefs, and small kelp farmers. The overall vibe was that the general public needs to get jazzed about Kelp, and fast, because we can use it in a multitude of ways. We love things when they're useful to us. Utilitarianism is such a hot way to sell interest.

I'm sorry Kelp, that we find it hard to love you without knowing how useful you are to us. You don't need a rebrand. You don't need to be some new-fangled "zoodles" dish in order to justify your existence. You're great the way you are, swaying in the current, traveling upwards towards the sun.

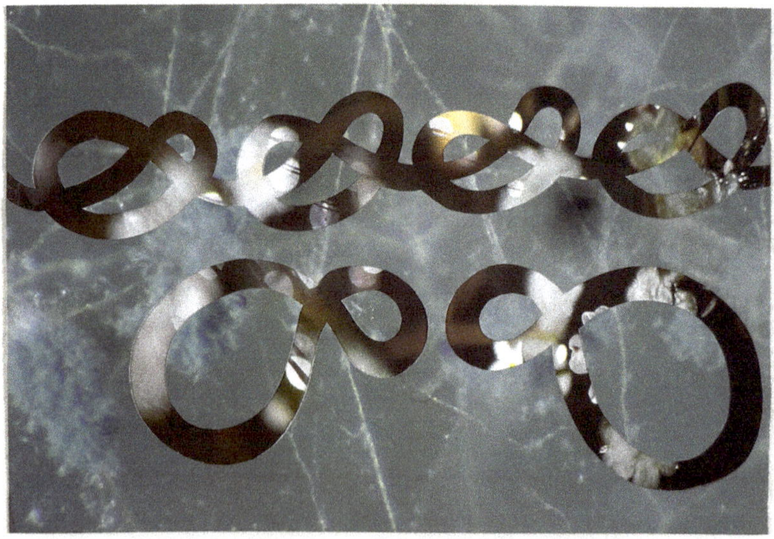

Image 45. *Knotted Light* by Rochelle Voyles

Emma Stanfield *(PhD Student)*:
Stanfield is grateful to be working with bull kelp as the study system for her PhD. She grew up on the coast of California, exploring beaches and washed up kelp, but only in the last few years has begun to understand kelp and its ecosystem scientifically.

My apologies to kelp come in many forms
And probably not enough
At the low tide I can confess my most calm and heartfelt emotions

My sing-song whispers of condolences and apologies
Soprano tunes of awe and alto admiration
Do you catch them the same way you harness offshore waves into your belly?

I clamber from your holdfast, along your guiding stipes, towards your alien head full of noxious gas that threaten to kill me

I may settle in, comb your thick sticky blades like hair,
Notice your bumps and tears and huddled epibionts,
Taking care not to puncture tissue or disturb bright eyed patches, young of the year

The rest of our day together is transactional, a cycle of stick-poke-swab.
I try to hold on to those holy moments from earlier, together, when I started my day in flattened beds above a preciously calm and graceful low tide
How can one extenuate our human actions – our own man-made gas and smoke that threaten to kill all life?
My own intentions are convoluted – both selfishly spiritual and scientific – causing me to cut, stomp, and sacrifice those bronzed kelp blades.

There are no algal-centric reparations for these individual or collective actions that can be stored into a document
There is no money I can pay you in return for your service

Yet an apology to kelp, for myself, I provide:
For the physical harm I have inflicted,
For brutalizing my privilege and desire to understand you,
For fixating you into the cold process of western science,
For not knowing you sooner, you and your grand and resilient force

Daniela Pierro *(PhD Student)*:
Pierro is a molecular ecologist studying host-microbe interactions across scales–from the molecular to ecological levels. She is especially interested in how symbiotic relationships may affect the adaptive potential and resilience of marine species, in the context of anthropogenic climate change.

The quietness of kelp is too often misheard as permission, for which apologies are long overdue.

My toes divot the sand in a plodding trail, occasionally obscured by black plumes of leaping sandfleas. Bright orange rays spear the innocent clouds, spilling vermillion across the horizon line. Salt crystals and rotting, purple berries litter the spaces between tangled heaps of algae, strewn along the beach that extends like a craggy finger before me.

The back of my heart twists and my throat constricts as I see you – a decapitated bulb of bull kelp, languishing on the rocks. Kneeling, I lift you, and the setting sun illuminates a reflection of my eyes in the orb of your head, still so slick and lifelike. We see ourselves and each other in an instant that lasts forever. The final fleck of day slips into her dark, roaring blanket, and our eyes are gone.

A genuine apology demands more than just a dialectic aimed at finding some truth across human and algal tongues; in the spirit of reciprocity, it also includes the metaphysical task of developing trust, and first learning how to, by listening.

The sloshing of seawater trapped in our bloated necks,
the tension of a full breath held in our spherical faces, the

rustling of thin blades that cascade in soft, glossy curls and brush against waddling crabs and now, your slimy, reverent fingertips. Hear our sunbaked blades lightening to green and tan and sighing against the salt breeze until we flake away like an organic scattering of ashes.

What do *you* think is the best path forward? What mode of apology is the most meaningful to you?
When we are embarrassed or uncomfortable saying something directly, people tend to sing, whisper, or palliate our voices. Absentminded in the kitchen, *why can't I find the pepper?* You look *am-a-zing!* I feel sad *all the damn time.*

Deepening to a serious tone centers the full heft of our opinion, while weaving in a little melody does the opposite—we hover above this weight, avoiding ownership over the sentiment of our sentences. Switching to another register, especially a high one, can rescue us from the gravity of the bare spoken word.

Why is it that when we are ambivalent, we fly to these escapist, trivializing solutions? Is song the least appropriate way to address the saccharine, salty body, a bubble trapped in her own throat? What would it mean, to sing an apology, and what is the possible scope of a song?

Then, I remember Henry in my arms, and the lullabies I offered his wide puddle eyes and perfect toothless smile that my entire being is utterly obsessed with. He demands this time with screams and wriggling that only calm after white sheep, *white sheep on a blue hill,* my whimsical blabbering of unpasteurized affection.

But musical expression is not confined to this narrow space of awkward glossing-over or infant tunes. It takes breath and energy and feet grounded in the sand and a vulnerable, irrational part of the soul that might even be uniquely fitting. It invites us to harmonize and form new motifs together. It loves and it soothes every crevice that comes between us.

So, stretched horizontally on the sand, I wait, and I listen for you. The faint buzz of sandfleas, the piercing screech of a territorial oystercatcher, the call of a distant eagle.

Beyond this racket, the delicate web of my holdfast clings to an old stone in the cold seabed–a pop and crunch as it dislodges, and my gurgling journey to the surface begins. The fullness of the ocean reverberates around the gentle rock of my air-filled stipe and the swoosh of my long wiggly blades that grace the scales of passing fish, amid the slowly growing, now deafening, sucking of waves at the sandy, pebbly surface the two of us share.

Knowing I will never receive your forgiveness, and knowing my apologies will never be enough, can I still live my life in a way that soothes our intertwined histories, that begins to heal our ongoing relationship? Is there a way to create this future together?

A thud on the beach. Squeezing, hissing from my pouch of air, the smack of a wave against my hollow body, now sprawled at your side.

Slow and lacrimosa, the melody is flowing over us, and I realize it is coming from my own breath. May I cradle you in a soothing bath of sound? Ad libitum and following your cues, can I offer you a harmony, to marry and to magnify your exhalations? In this new music that combines our voices, can we one day flourish mutually?

Becs & Max Apologies

03/12/2023 Becs

I apologize to Kelp for not understanding it. I think Max has explained your reproductive process like three times and I still am a little confused.

03/26/23 Max

I was thinking about apologies from Cathy Busby's book last night, and realized I didn't take enough pictures to understand her full argument about what makes an apology 'sincere' or valuable. But from what I did have, I felt like my apologies will come from 'collective' and 'personal' categories she describes.

From the collective: I feel sorry, in my capacity as an individual representing a larger social entity.

From the personal: I feel sorry for my own completely private transgressions and shortcomings in the interest of kelp, despite (maybe especially because?) the time spent trying to study it.

I'm sorry that we prioritize a completely made up entity called the economy over you, and I'm sorry that we have divorced 'economy' from 'ecosystem.'

I'm sorry that I still don't have a good recollection of the texture and smell of kelp blades, as I've only really been on kelp territory once or twice. And I'm sorry that I thought your texture was gross and slimy for so long - it may have been part of a general aversion/rejection of unfamiliar things

03/26/2023 Becs

I am thinking about the video I watched from Josie and of how it brought us from the surface down into the ocean and back up into the air; it made me feel for the creatures underneath the same way as I feel on a hike. It was something those of us who do not dive deep into the water never get to experience. How can I apologize when I do not know kelp in a haptic way? That video can trick (in a good way) my brain into sensing that touch, of my body as the kelp sways. I am sorry to kelp that I am still trying to relate to it, if it even cares about that. I am sorry that I don't like the furikake seaweed flavor I have in my house, so that corpse is just existing in minor dislike, instead of being treasured or being free.

I am glad I am trying my best to make you, kelp, cozy in your lab home through Max and my clay experiments.

03/27/23 Max

I felt colder than I have felt in a while today. I'm sorry for my part in the processes that have changed the temperature of your home - I know you have evolved in the cold, and that you, bull kelp, can only live healthy full life cycles in the cold. I know you don't really have anywhere else to go - maybe just north? I'm sorry for the ceaseless, worsening stress of living at a temperature you weren't born to live in. Even if you can overcome it, and I hope that you or at least your descendents do.

04/02/2023 Max

I'm sorry, to the kelps in my lab, that I haven't been in to check on you in several days! I have not forgotten you, and I will be in more next week.

05/10/2023 Max

I'm sorry to the kelps in my lab that it's taking us a long time to build up a comprehensive husbandry protocol.

5/12/2023 Max

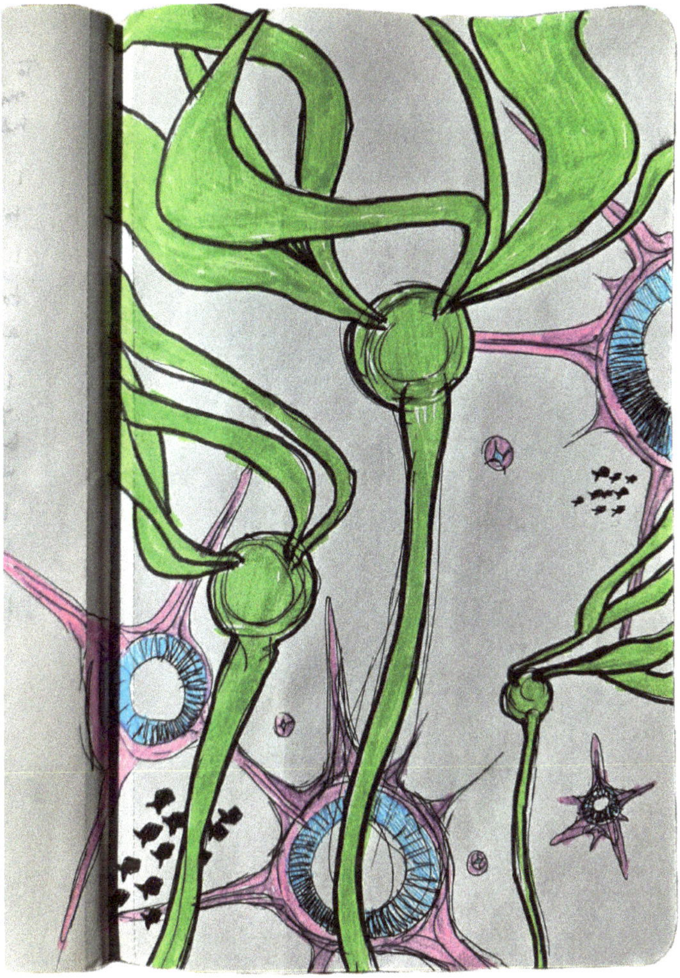

Image 46. Sketch by Max

05/14/2023 Becs

Each time I talk with Max, I feel like I understand kelp a little better. Kelp and I are still in the courting phase where my footsteps feel a bit unsteady. Kelp has existed for millions of years and to it, I am just a baby teething. In reference to humanity, we are that same baby but with a super powerful flame throwing blasting away. I am sorry, specifically, to the kelp in Max's lab on the high silica high firing substrate. They aren't growing as well, or sometimes, at all. I am sorry that you are collateral damage in following scientific inquiry. I am sorry that while I was able to give you a bed of clay, it is in some way more inhospitable than the plastic base of the petri dish.

6/21/23 Max
my kelp babies are fucking dying
i'm sorry

07/12/23 Becs

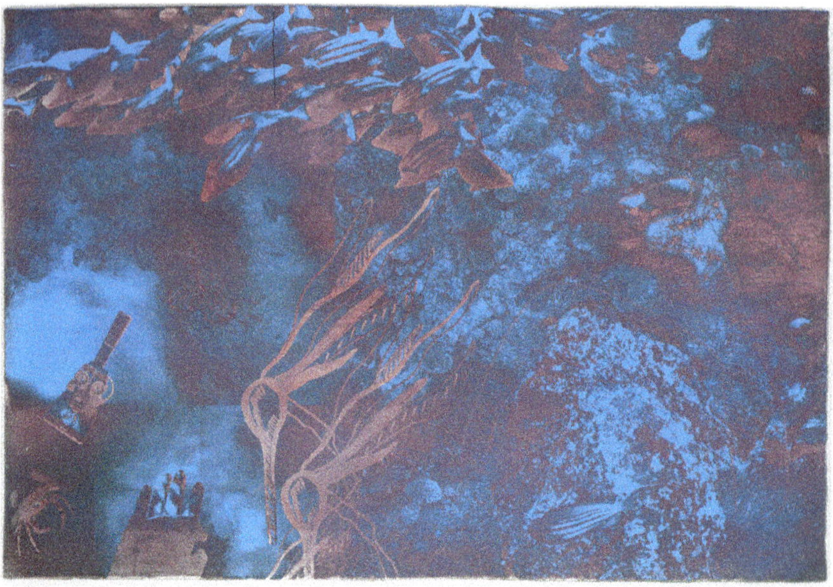

Image 47. *Beneath the Tide* by Becs

11/09/23 Max
I'm sorry to six species of kelp in particular: *Nereocystis luetkeana*, *Hedophyllum nigripes*, *Alaria nana*, *Pleurophycus gardneri*, *Laminaria setchellii*, and *Postelsia palmaeformis*. I took tissue from each of you - more than two grams dry weight - for an experiment. That tissue told me things nobody else knew before, but I didn't need it more than you did. I don't know if clipping your blades 'hurts' you the way cutting my skin hurts me. I just had knee surgery though, and I feel sure that the cells in my leg were screaming 'STOP' even though my brain was not awake to hear them.

11/14/23 Max

Dear kelp in my lab this summer,

I'm sorry we kept you in petri dishes and flasks. I learned well after I had already moved you into a 2L flask that you may need dozens of gallons to grow to your 'natural' size. Constraining your growth was not my intention.

11/14/23 Becs

I am sorry that I don't think of you everyday. Sometimes you slip out of my mind like leaves on a river, and I want to be with you in the ocean, feeling the tide around me but sticking in my holdfast, solid and with friends. When I close my eyes, I can see your blades in the deep water, sun filtering down onto us, with fish flitting between. But, but, I have only been scuba diving once and I didn't spend much, if any, time with you. I was mostly worrying if my ears would rupture because my nose piercing was stopping me from equalizing. I didn't pay attention to you; instead imagining blood seeping out of me. I am sorry for being so physically close to you and not practicing kinship. I know more now, and I really am trying. And I am still sorry. I hope to someday soon be a better comrade to you. For now, I give you my heartfelt apology.

11/20/23 Max

Dear kelp, I'm sorry - I was cynical and defeatist this week. I felt angry with a lot of things - most of them political or abstract or outside of my control - and slipped into thinking about how none of my kelp efforts really matter (or at least they won't matter soon enough). It was a myopic, and kind of self-absorbed flavor of nihilism that I am trying to move away from.

11/20/23 Becs

I wonder if you are tired. I'm so exhausted. How do you rest as your home shifts so far from how you have known it? I can barely process all the change whirling around me. I feel like my sense of self has constantly lately broken into a million pieces; I keep trying to tape it all back together. Can you teach

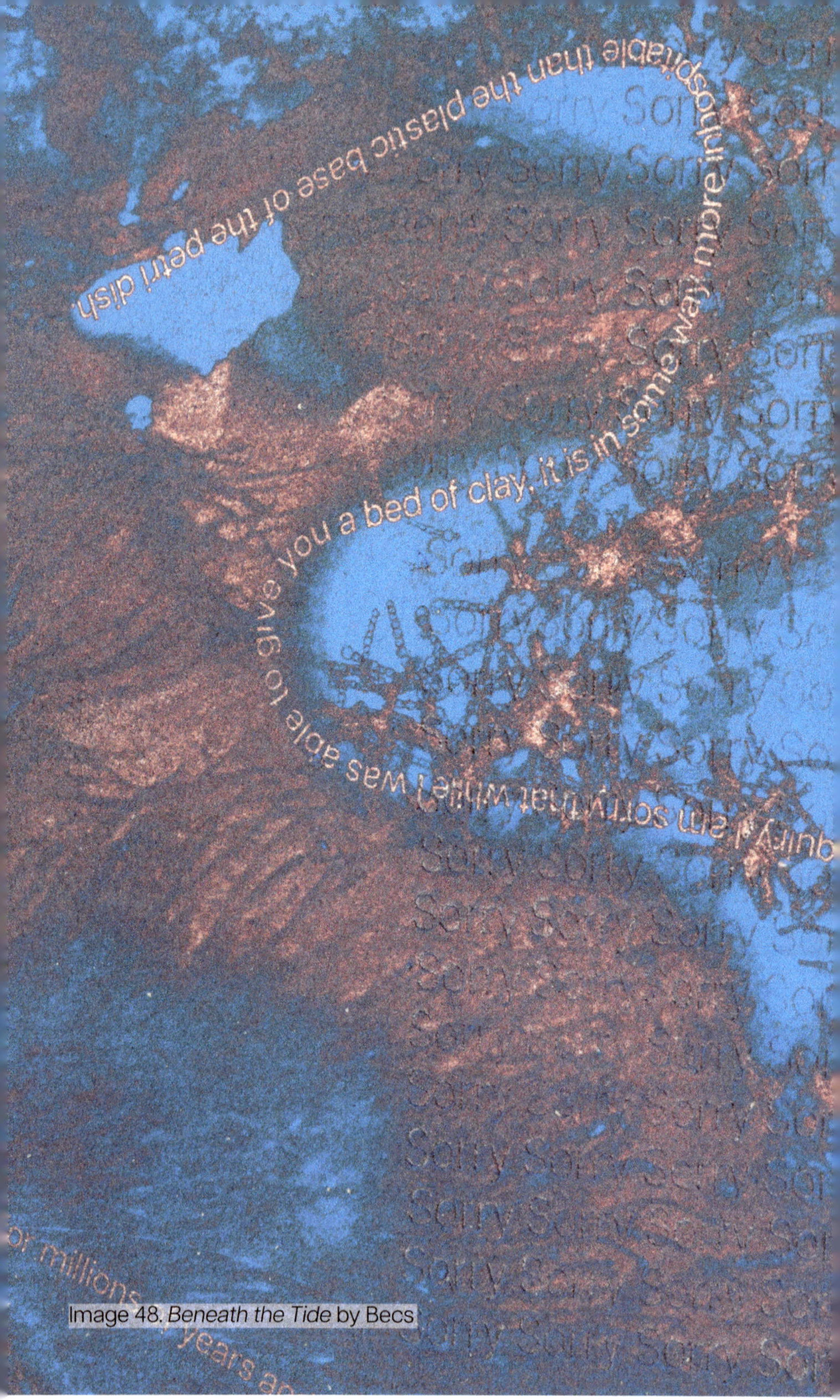

Image 48. *Beneath the Tide* by Becs

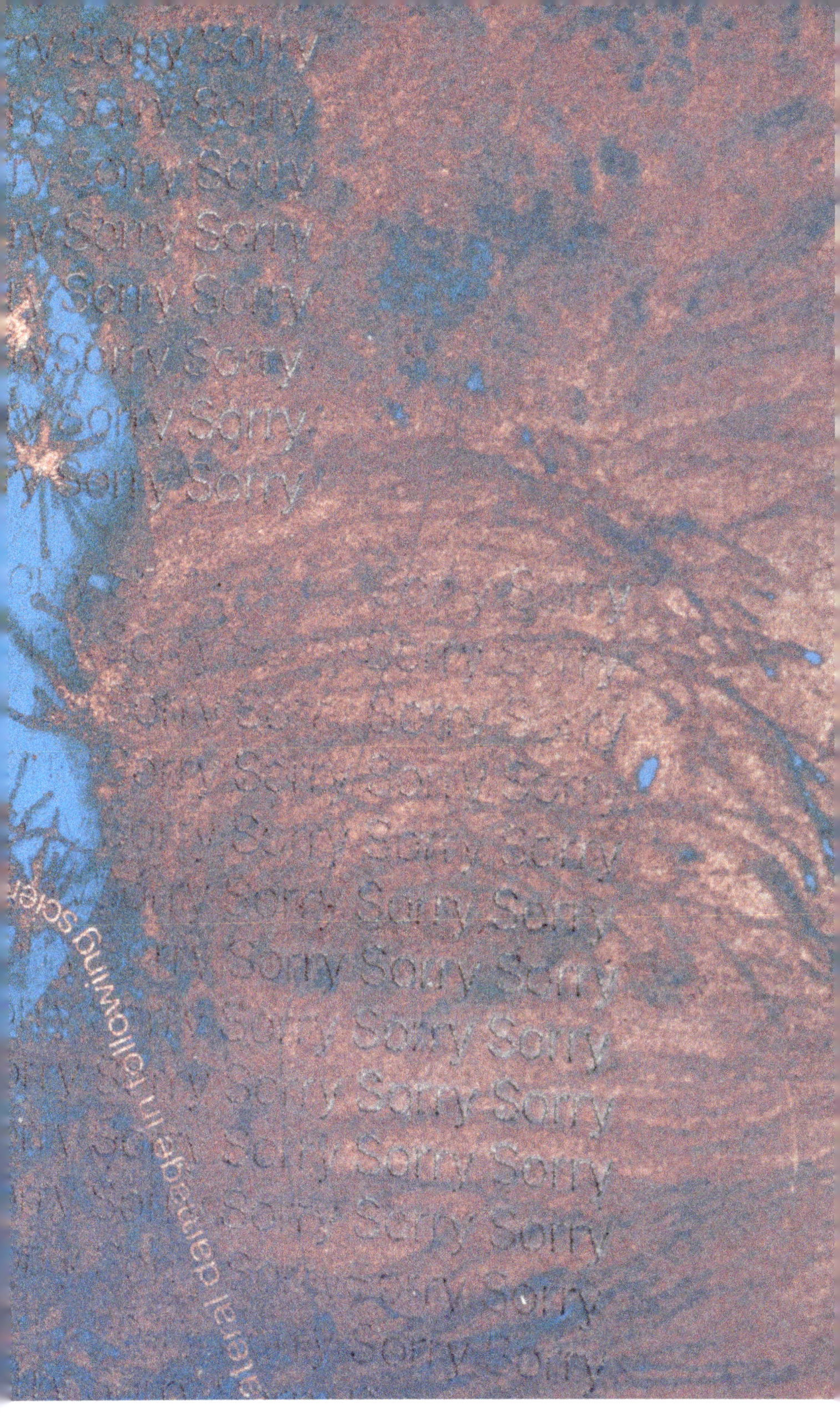
Sorry Sorry
Sorry Sorry
Sorry Sorry
Sorry Sorry
Sorry Sorry
Sorry Sorry
Sorry Sorry
Sorry Sorry
Sorry Sorry
Sorry Sorry
Sorry Sorry
Sorry Sorry
Sorry Sorry
Sorry Sorry
Sorry Sorry
Sorry Sorry
Sorry Sorry
Sorry Sorry
Sorry Sorry
Sorry Sorry
lateral damage in following

me how you stay strong? I am so sorry that human actions have set your world ablaze. I won't give up on being a support system for you, I won't give up caring for you, Max and I won't give up on saying sorry to you. I am sorry that I am so far from you, though you probably don't care. That's okay. Just know that I am sorry.

11/28/23 Max
I'm sorry, kelp, for describing you imprecisely - frequently. When I tell a new human what I do for a living, it is so much easier to say you are 'like a plant' than to say your ancestor endocytosed a relative of the ancestor of all land plants. I'm not sure it makes me any more understandable in the long run, and it saves me only a few seconds...

11/28/23 Becs
This past week I saw a painting that reminded me of you. It was at the National Portrait Gallery in DC, in the Forces of Nature: Voices that Shaped Environmentalism exhibit. The painting is Julie Packard by Hope Gangloff; Packard is standing in front of a background of towering kelp idyllic fish. I am putting so much attention onto this painting because Packard helped found the Monterey Bay Aquarium and is the current executive director. I'm sure you know this, but Monterey has incredible people studying and trying to rehabilitate kelp. So in a way, I see those scientists as constantly apologizing and Packard as the catalyst for us understanding you enough to attempt heartfelt apologies. I am sorry that so little has gotten better for you since Packard founded the aquarium in 1970; 53 years later and I am truly sorry that you still need our help.

12/5/23 Max
I'm sorry for my wastefulness this week - for multiple disposable plastic cups, utensils, and take-out containers, for running my laptop nearly into the ground because I couldn't figure out how to run the job more efficiently so I just did it sloppily (which will of course lead to a much higher impact form of consumption when it does finally die and I buy a

new one), for taking Lyfts instead of the bus. I'm sorry for the indirect harms that these actions may cause you - I have seen the plastic flotsam clogging cold rocky beaches, and I've heard that companies are now strip mining the ocean floor for cobalt to feed the electronics industry. I'm sorry.

12/05/23 Becs
I just tried writing my first thank you to kelp and well...it was so focused on me. But maybe that is part of what thanking is for? To make sure you know how much you mean to me? I'm sorry that I am asking you to tell me how to support you. I am sorry that today I imagined getting a piece of you and laser cutting shapes out to act as bookmarks. If you were fresh, you might ooze, I'd have to ask Max about that. I've seen you more lately dried in the bright lights of Joong Boo Market. I can almost smell the fishy, burning smell of the laser vaporizing you at my press of a big red button. It's a more dramatic way to face your end than slow acidification in the sea. But that doesn't make it okay. I am sorry that I daydreamed of your corpse today.

12/13/2023 Max:
Dear kelp in Puget Sound,
I'm sorry you are suffering. I spoke to a friend today - who also researches kelp - and I worry I've become blasé (or at best clinical) when thinking about the health discrepancy between you and kelp on the outer coast. You have poorer health, and depleted microbiomes; you absorb less nitrogen than kelp from less polluted waters. If I were to describe another human in such dire straits I would probably say they were starving.

12/13/2023 Becs:
I am sorry that I am taking so many planes over the holidays which must contribute to emissions which tip the scales to your demise. I have family all across the country and love seeing new things, and if I saw your carcass in front of me as a direct reaction I know I would want to change, but I'm

not sure if I would. See, the thing is, I am selfish sometimes. And at the same time, I care for you, deeply. I am sorry that I have not found a balance between my wants and your needs. I don't eat meat because I can imagine the cow in so much pain as well as baby cows asking for pets just like our household pets. Dogs will survive climate chaos though they will suffer. I'm not sure if you will survive. I really hope you do. I am sorry that today all I can send is holding you in the light and wishing you my best.

12/15/23 Max:

Ode to Nereocystis
I want to build webs, like you do. Yes, also like a spider. Or a fungus. Or, I suppose, a neuron.
I want to stretch out my blades and bring about great syntheses.

How shall we compare?
The way you fold a megamillion microbes into your mucus, into your lifecycle.
I would take more things in stride, into account. What demands the world has of me, I would answer, like you answer for the nutrients and needs of your neighbors.

The support you give - the shelter - for others to spin webs of their own: for otters to cross wires with urchins, for beach dwellers to cohabitate safely with the tides, for your genes to interact with the genes of your microbiome, and for iodine to change and be changed by the troposphere.
I would be so steadfast, so reliable; a node in my own web, stably holding tension and passing along vibrations to the network around me.

Nereocystis, can we be co-conspirators? Collaborators in forging connections? I would draw a map (portrait) of all your multidimensional interfaces, to capture your expression and represent you as faithfully as I know how.

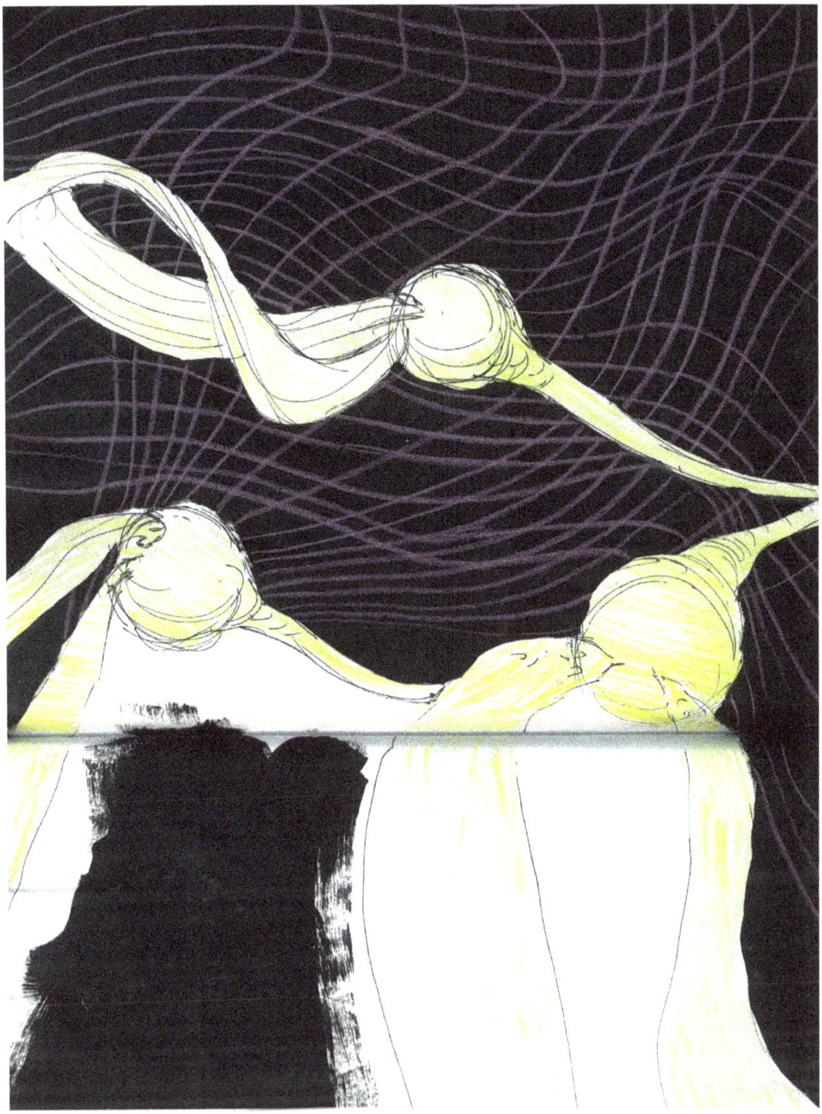

Image 49. Sketch by Max

12/20/2023 Becs:
I am sorry today that I don't have my full strength of brain power to attempt to help/understand you. I feel a bit in a blur with rough edges. I am sorry.

12/24/2023 Max:
Pacific coast kelp, I'm sorry for wishing the rain away and

secretly being excited that it's been such a mild winter. It's easy to say, abstractly/theoretically, that I'm ready to sacrifice more sunny warm days in favor of a stabilized climate. But I also crave the dry, hot climate that is so harmful to you. Clearly it's going to take a while to internalize the ideas that Becs and I have talked so much about - that you and I are in relation, and your needs should not be subordinate to mine. In the meantime, I'm sorry. I'll try not to act on my warm-weather cravings.

12/24/2023 Becs:
I am sorry that your lifecycle still confuses me. Max and I decided that I will make a new graphic of your lifecycle, and I am just sure I will get so many things wrong. I'll get advice/knowledge from Max, and yet I still see this imagined future of every draft having one thing (even if it is minor) incorrect. I'll try to do you justice.

12/28/2023 Max:
Dear kelp, I'm planning to get a gift for a family member, and I know I'll be spending money with a company that contributes to large scale waste and pollution. I'm sorry.

01/06/2024 Max:
I'm still thinking about the kelp Becs and I grew on substrates. They hung on for so long, but when I found contamination I decided to call it, which involved bleaching the container they grew in before pouring everything down the sink. I'm so sorry for killing you with bleach. I felt terrible - I've felt a bleach burn before myself.

01/06/2024 Becs
Dear kelp,
Hello there. It feels like it's been a while since we last spoke. I've missed you. It has been about two weeks since we spent theoretical time together. I am sorry that I step away from you at my convenience. Sorry again.

01/16/24 Max:
Dear kelp microbes,
Hello, and I'm sorry I've never addressed you before. I'm also sorry for treating you like research objects. Those of you who grow up in lab are very unlucky, I'm sorry you won't get to reproduce and evolve as you would have in your natural setting.

01/16/24 Becs:
Dear kelp,
Sunday I saw Mean Girls with a really fun group of new friends. We were all updating each other on life and I told them about working with you. And it made me feel so warm and excited, and yet...I don't think I explained you well enough. I tried to make sure they knew how important you are for the overall ecosystem as well as your inherent worth as a living being. My only excuse is that it is pretty hard to explain after two cocktails and a dinner of movie popcorn. With that measly excuse made, I am sorry that I try to elevator pitch explain your plight and your triumphs. I am sorry that I made you into a sound bite of my achievements; said in the same breath as my exhibitions in the spring. You are deserving of space to expand, contract, and exist on your own terms. I am sorry for putting you into a humanist box.

01/23/2024 Max:
Dear kelp, I'm sorry that this week I overpromised and underdelivered. I just get so excited about the possibilities; the future potential of good ideas and things Becs and I want to do to understand you. I just want to say that I DO want to realize every plan and thought we have to make amends! I want to, so this week I said I would, but I fell behind. I've done it before but this week it was extremely avoidable. Sorry.

01/23/2024 Becs:
I am sorry that I only sometimes recycle now. The thing is Chicago barely recycles! So it just seems pointless when I know with almost 100% certainty that you will just be thrown

in the trash. I know this may seem off topic from you, but let me explain, please. In our world where everything is so interconnected, I see trash as harmful to you; I hope that it is not presumptuous to act like I know who your enemies are. Well, what if the "recycling" (which I did not recycle) falls off the truck and then gets swept away into the water! (though I guess we are very far from your habitat here on Lake Michigan) What if the additional carbon dioxide given off from landfills just makes the ocean warmer, and warmer, and it hurts you? I am trying to reuse items as much as possible with the inaccessible recycling and yet, I am sorry for not doing more.

02/06/24 Max:
I started reading *The Expanse* series this week, and it made me think of you. The series is a science fiction space adventure - it's focused on humans, and how they deal with/traverse the incomprehensible scale of outer space. This pretty naturally includes a focus on the expansion of the human race: it seems that most SF futures can't be imagined without human beings colonizing 'new' lands. I'm sorry that even our most speculative fictional imaginings are tinged with such an assumption about the relation between humans and non-humans, and I'm sorry because I'm sure I also make that assumption in many contexts. I hope human stories will grow increasingly collaborative. I hope the narrative of conquest loses some of its power.

02/06/24 Becs:
I apologize for, well I do not know if you need this apology, but I feel like I've taken advantage of you. I got accepted to this incredible opportunity at Friday Harbor Labs which is really exciting, and at the same time, I got accepted based on a proposal where I talked about the project with you and Max. I did not propose that as my artist proposal for the residency, just so you know. I sort of flexed my art/science combo credentials. Those credentials are, in large part, because of you. And that is lovely! And yet...I don't want to be dropping

your name to gain respect and support. I want to give you respect and support.

02/13/2024 Max:
Dear kelp gametophytes in lab,
I'm sorry, it's been a while since I visited or talked about you guys. I think we should try adding clay substrates back into your environment - it seemed like you did better with them than you have been without.

02/13/2024 Becs:
I am sorry that I keep buying water bottles which only adds to the destruction of the environment. I don't have a water filter at my apartment or at my art studio and drinking the straight tap water just grosses me out (you might understand if you saw the sink at the studio). I have been putting off spending the money on a new filter. Honestly, I am constantly dehydrated lately between buying gallons of water. So for my own mental health it feels needed, and yet, everytime I feel a pang of guilt in the pit of my stomach. I am sorry that I do not plan on stopping yet. We both need water and I am sorry that my consumption negatively affects yours.

02/27/24 Max:
Dear kelp, I'm sorry but this week I've thought of you mostly as the small cluster of pixels that represents you on my laptop screen. You're an event in my google calendar, an icon in a slide I'm preparing, a note jotted in the margins of a PDF. I know this does not reflect your reality at all, and I will try (especially this summer) to build the habit of imagining you more completely.

02/27/24 Becs:
I'm sorry that I have not recently added to my understanding of you. I have been focused on understanding the perspective of scientific thought versus artist thought and finding what way I can distill the origins of Humanism which I learned in Sara Black's class, Beyond Object, Past Subject.

03/10/24 Max:
It's getting close to the Nereocystis reproductive season - every spring, microscopic bull kelp babies are expelled from their mother's sori patches and float through the turbulent waters of the intertidal. When they reach a stable substrate, they dig their holdfasts in and start their year-long growth to maturity. I'm sorry to the spores that won't find anywhere to root, and sorry to the spores that are successful for any stress from the journey. It seems like it would be scary, being micrometers long, but wafted along in currents that are measured in kilometers (9 orders of magnitude difference in scale, I think?).

03/10/24 Becs:
I am sorry that I did not intercede in a massive amount of plastic trash being made this week. I was with my cousin in the hospital for three days...and she is really susceptible to infection right now, so pretty much every medical object had to be thrown out after one use. A few things could last for a few days but then they just were not safe for her anymore. Containers of saline water, syringes, alcohol pads, endless food packaging, medication vials, ice packs, medical tape, and honestly more than I know the names. She needs all of this equipment to try to recover so I do not regret it, and at the same time, I am sorry that so much trash is created. I know plastic waste can really affect you. I am sorry.

03/12/24 Max:
I'm sorry about the amount of waste I produced during my experiments today. I went through like 20 plastic cuvettes just to get a few confirmatory readings of cell density. It's 'good science' but it's just a lot of plastic that we toss or put toxic chemicals all over before tossing. I'm expecting a lot more of the same if I want to have a successful project going forward, but maybe I can streamline some of the steps when I scale up? I will be thinking of you. Sorry.

03/12/24 Becs:
I am sorry that Western theorists made our relationship into collateral damage. Locke was trying to give the people a way to grasp power back from the monarchy and somehow created a whole new hierarchy that cemented humans as the only ones with power. I can feel that philosophy echoing back when I am trying to be your kin.

03/27/24 Max:
I'm sorry for my research schemes - in the past couple of weeks I have arranged to have you freeze dried, ground up, and chemically extracted. I'm sorry to your relatives for the jet fuel and gasoline I'll burn on my way to get you.

03/27/24 Becs:
I am sorry that I am unsure of how to help you, how to make it clear how sorry I am. It has been a hard day and I wish I could be what your holdfast tethers to. I am sorry that I am not that today. I am sorry that I live in the midwest and to have any chance of physically being with you, I will need to use massive non-renewable resources.

04/02/24 Max:
Dear kelp, as an inhabitant of the oceans i.e. our major plastic dumping grounds, I'm sorry about how much plastic I used today.

04/02/24 Becs:
Hi kelp,
I'm feeling a bit disconnected from you. I'm sorry for letting my focus drift away from you and onto theory and history.

04/09/24 Max:
Dear kelp,
Sorry for the news cycle. The eclipse was magnificent, but I also started to think it might be a convenient media fest that helps us ignore more pressing matter. I wish you got more press time - I think you deserve it at least as much as syzygy.

04/09/24 Becs:
I am sorry that I am really hungry for dinner and I feel a bit distracted. I am imagining you in front of me, towering above me. A giant of the sea. I can taste the salt water in my mouth, though don't you adapt well to fresh water as well? I am sorry that even my daydreams of you
are an idealized, movie version of you.

04/16/24 Max:
I'm sorry I've been out of sync/out of touch. I'm working with your proteins and your microbes and your gene sequences, but not actually with you much these days. I'll see you soon though - I'm making plans to seek you out. Please meet me in May!

04/16/24 Becs:
I am sorry that sometimes I treat you as a collaborator and then, once in a while, I treat you as a muse. I want us to be colleagues and not to replicate a human-oriented hierarchy. I am sorry.

05/14/24 Max:
To several kelp species in Monterey Bay, I'm sorry I missed you on my trip last week - I didn't see you, touch your blades, or wait for the right tide to go visit you. I meant to get closer, and to spend more time with you, but I was rushed and pretty brusque instead.

05/14/24 Becs:
I am sorry for not greeting you yet. I think I saw you under the surface last night, but I wasn't sure. Eric was pointing out other creatures and I felt nervous to interrupt.

05/22/24 Max:
My apology today is a little hollow - it is not specific to you, you are instead one intended recipient out of multiple - but I want to be representative so here she is. I'm sorry, in general,

this week. I've been caught up in the news cycle, especially news of the humanitarian crisis in Palestine, but, but it never feels 'caught up enough,' and it sometimes feels absurd to make apologies to you, a non-human, when I hear of tens of thousands of humans being killed on time-scale that's even faster than you and your species being killed. I feel sorry in a kind of panicked, impotent way - sorry to you, and the people in Rafah, for disruption/destruction/death with no clear end date.

05/22/24 Becs:
Hi bull kelp, I ate a bit of you this week and I am sorry, not only did I eat a bit of you, I shared you with a new friend Cricket, and then, yes, I know, there is even more to this. I took you home with me. You had already been banged around and snacked on in the water so I thought why not. But I am sorry. I do not know what to do with your carcass. I am sorry for hurting you like this. I feel closer to you after feeling the crispness of your blade mash between my teeth. I am sorry.

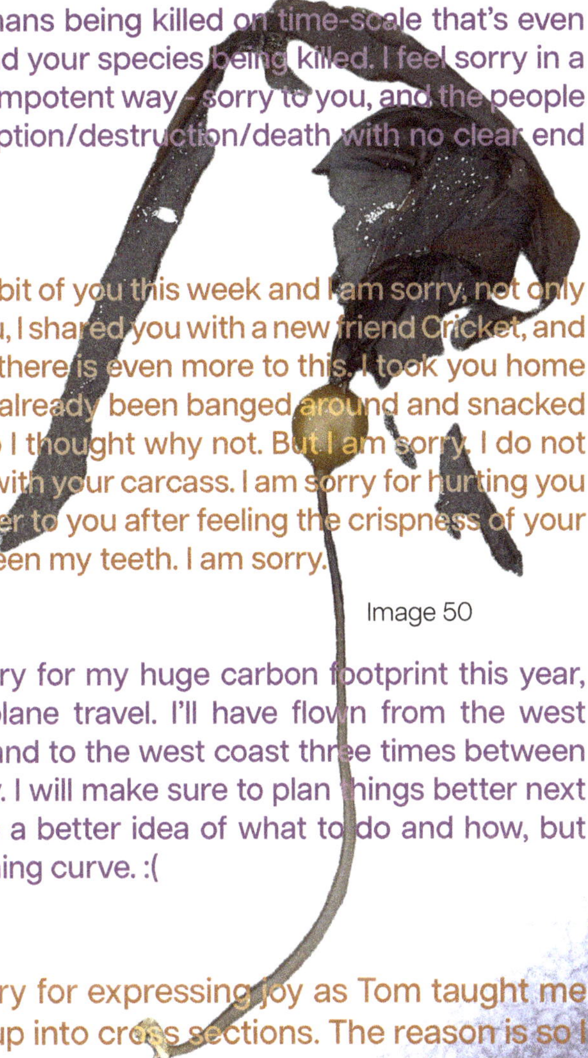

Image 50

5/31/24 Max:
Dear kelp, I'm sorry for my huge carbon footprint this year, especially from plane travel. I'll have flown from the west coast four times and to the west coast three times between January and July. I will make sure to plan things better next year - now I have a better idea of what to do and how, but sorry for the learning curve. :(

5/31/24 Becs:
Hi Kelp, I am sorry for expressing joy as Tom taught me how to cut you up into cross sections. The reason is so I can look at your insides in a microscope, gory right?

It feels like magic, the way light travels through your cells.

My bus driver yesterday told me about how there used to be giant swaths of you all around the island's shore

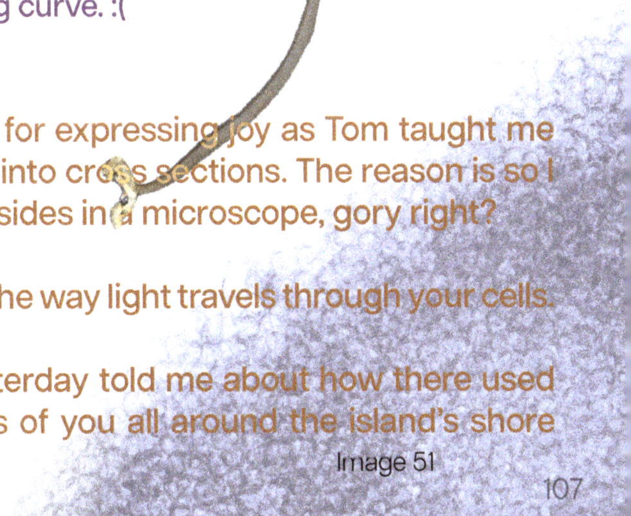

Image 51

and how that has all changed within her lifetime! As a teenager, she would boil clams or mussels in sea water on the beach as a casual fun day. Now those same sea creatures are so difficult to find. And all of your stories are so interconnected within the intertidal.

I'm sorry that you are having such a hard time.

06/04/2024 Max:
I'm sorry to the *Egregia* I collected in May - for cutting your blade, and putting your flesh in a bottle deprived of light and fresh oxygen. And then trying to chemically strip your microbiome. I went to look for cicadas with a friend and labmate yesterday, and I found one that was partially eclosed - still white and emerging. We collected it (it died), and I keep wishing I hadn't pointed it out/taken part in the act of collecting.

06/04/2024 Becs:
Hi, I am sorry bull kelp, I found a small one of you washed ashore when I was tidepooling. I have to admit that I took you home to press you. I am not sure if you could have reattached and grown more but either way I took you from your home and for that I am sorry. I am also sorry that I plan to collect more of you. I think I am going to make a color coded map to figure out where collection is permitted. I am literally preplanning stealing you away and I really am sorry. I want to try to find a way to give something back to your brethren, maybe it will be blood, maybe it will be a trade of micro bacteria through a kiss of gargling.

06/15/24 Max:
Yesterday, I apologized by trying to exchange microbes with kelp. Becs and I have talked several times about the physical ways we could make amends to kelp; the substrate experiment was one major way we thought of, but we also discussed giving parts of our own bodies, in honor of the samples and clippings we've taken from kelp thalli. We

considered scattering a few drops of blood into the water near kelp - adding precious iron, and maybe some useful carbon or nitrogen to the environment... But yesterday, it was a handful of seawater that I swirled around in my mouth for a few moments, then spat back onto a kelp blade.
I don't know if the microbes from my mouth can provide any useful resources to kelp. In the short term of course, there's some extra nutrients. But in the long term, on evolutionary time scales, encoded by fixed changes in the DNA? The chances of a useful exchange are vanishingly small. Advantageous mutation or horizontal gene transfer between cells likely requires thousands, millions, or billions of moments of interaction and exposure. From that perspective, my mouthful of seawater probably represents the tiniest of opportunities for trade, almost nothing really. But maybe Friday, June 14th around 4 or 5pm was an especially serendipitous time. It's a total numbers game, but I did it with the idea that that might be true.

07/14/24 Becs:
Hi kelp, specifically bull kelp
I am sorry for biting you. I told the story to Bellamy this week of tasting your blade and realizing how sweet you are. It is a funny story and I am thankful that I had that experience with you, and at the same time, I feel this deep sensation of regret for this new way of interacting with you. So I want to say again, that I am sorry. I do not think I deserve forgiveness, and I also accept how that choice benefited me. It helped me feel closer to you and I will still always be sorry.

07/14/24 Max:
Dear kelp,
I'm sorry you're going through the heat of this year. It's not terrible in Chicago, but my parents in Oregon said the temperatures were well over 100 F - which is not how I typically think of the Pacific Northwest. I don't have any reason to think this would make you "feel" better, but the heat's hurting us (humans, US Americans) too. I think I heard

reports that the Colorado River is becoming an increasingly controversial topic in the southwest US because too many different cities want (need?) to tap into it for water. I'm sorry, and I hope you'll be OK.

07/30/24 Becs:
Hey, I wanted to say sorry to you today. I spent a lot of the day pounding the image of your body into clay. I am making ceramic panels focusing on intertidal creatures I met at Friday Harbor Lab. So I laser engraved wooden stamps of close up images of your cells, pushed one hundred pounds of clay past Lollapalooza set-up, rolled out slabs of clay, shaped them, and they hammered (actually with a jar of underglaze) the stamp of you into the leather hard clay. It looks really amazing and at the same time I am sorry for this additional form of violence. Also I am sorry that I just got really excited when I realized bull kelp as glaze might be possible (ash glaze is what it sounds like, ceramic glaze made from the ashes of any plant, though you are not a plant, so maybe it is not even possible, anyway). I'm sorry.

07/30/24 Max:
Dear kelp, I'm sorry I ground up your tissues and didn't use them. Well, didn't use them "in time." I compared freshly ground tissue to year-old ground tissue in an experiment last week, and realized there is a noticeable difference between the results from the two types. This means that I probably won't use the year-old tissue in any future experiments. Maybe I'll find something else to do with it, but right now it seems like that tissue will just go to waste. I'm sorry for over-collecting.

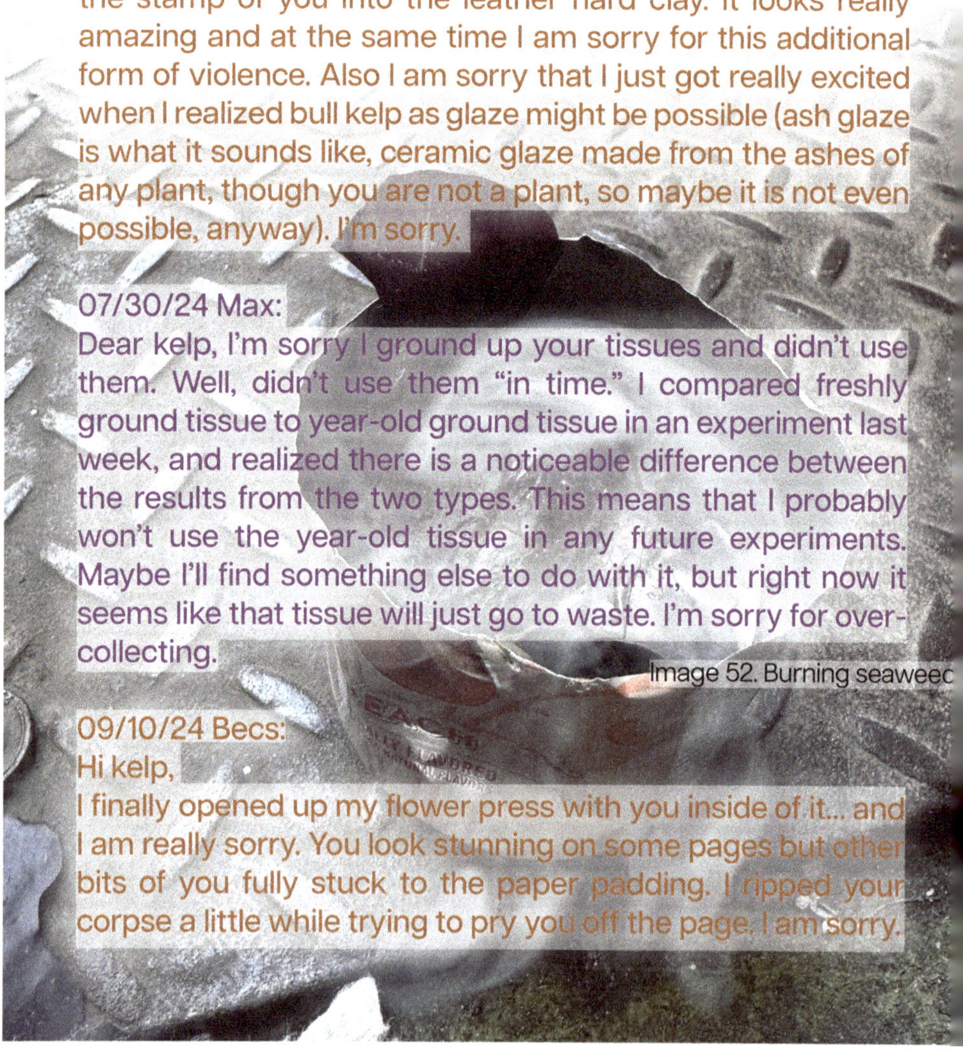
Image 52. Burning seaweec

09/10/24 Becs:
Hi kelp,
I finally opened up my flower press with you inside of it... and I am really sorry. You look stunning on some pages but other bits of you fully stuck to the paper padding. I ripped your corpse a little while trying to pry you off the page. I am sorry.

09/10/24 Max:
Dear kelp, I'm sorry, but I think mass harvesting and aquaculture is in your future. I love the thought of healthy, dense kelp beds and canopies—but I love it most for kelp that is wild and unrestricted. When we domesticate you, it will be ruthless.

09/29/24 Becs:
Hi. Today I need to say sorry to seaweed. I burned you (or I tried). I took the corpses I had pressed at FHL and put you in a sawed off La Croix can. As I crouched at the edge of the loading dock of my studio, I burnt my fingers as I tried to ignite your body. I turned you into a glaze of mottled blue-toned white with an edge of burnt orange. You were beautiful in life, and just as beautiful after I manipulated you into being preserved in glass. I am sorry for taking you. I am sorry for burning you. The smell of sweet musk is ingrained in me.

09/29/24 Max:
Dear kelp,
I'm working on a couple of manuscripts right now (hoping they will develop into publications one day), and the results don't quite match with my expectations... I think the data are telling us a story, but it's all still a bit muddled. It's not the crystal clear, mechanistic explanation that I 'wanted'. But that is OK - and I'm sorry I assumed things of you. I don't want to risk obscuring 'the truth' about you, because my own expectations were so rigid.

11/22/24 Becs:
I am sorry for almost throwing out your charred remains. I turned you into ash to use as a ceramic glaze and forgot what container I had put you in. Luckily, I remembered as my hand cradled a cut open La Croix can with the ashes. I will try to be more thoughtful.

12/06/24 Max: Kelp, I wonder if I'm missing something, in my reluctance to think of the kelp 'soul'. My life experience and professional training have not made much room for the idea of souls, either human or non-human. I don't disbelieve in souls, but they've never seemed particularly 'important' or unique to humans - and if souls don't matter, it's tempting to dismiss the idea of a kelp soul all together. Couldn't I remain true to a Posthuman perspective by considering human and kelp as equal in our lack-of-soulness? But increasingly it feels like I am losing out on a tool to connect with you, and to understand. I have tried to understand why kelp is the way it is and what allows kelp to do what it does through only the kelp body. Would we both be better served if I tried to answer these questions with the kelp soul in mind? What cosmic or nonrational forces drive kelp evolution and ecology? Are you 'trying' to make change in particular directions, and would the idea of a kelp soul help me see those directions?

12/06/24 Becs:
Hi Kelp, I am sorry that I bought a car. I know that each action I take which increases my usage of fossil fuels is another step into radically changing your habitat. And yet, even with that knowledge, I still am a selfish human. I am sorry.

12/13/24 Max:
Kelp - I'm sorry and not sorry about something. As in, I'd be sorry if it harmed you, but I don't think I will ever know that. I'm sorry for giving up on you a bit as a scientist. I'm applying for a job that would take me out of research, and into seaweed farming and sustainability efforts. I am hopeful about the new paths of understanding and communication that this choice opens up, and I think this path will be 'better' for me - but I have no guarantees that it will be 'good' for you.

12/13/24 Becs:
Today, I deleted Max's little name and date for truly no reason. We have turned who can write the titles fastest into a sort-of competition. I needed a laugh and was annoyed that they were so insanely fast. I may be getting a cold, just got my period, and did not sleep at all last night. I am sorry that my

unhinged exhaustion took even a moment away from our work of centering you.

2/7/25 Max:
I'm sorry for the contamination of your home. The seemingly innocuous, infinitesimally small industrial by-products and commercial compounds that leach into your chemical environment from the shore. I don't know how much of an impact these compounds make; I guess we'll all just have to find out as time goes on.

2/7/25 Becs:
This doesn't feel like my place to apologize, but I can't divorce myself from the actions of humans who theoretically represent me. I am sorry that the president is undoing environmental protections. I am sorry that you will suffer more because of him and his supporters. I am sorry that you are not who I think of first as new policy changes come out. I am sorry that as women , and people with uteruses, bodies are stripped of choice and control, I can't seem to hold equal space for you in my head. I am sorry that it feels like I can't protect myself or you.

2/14/25 Becs: I am sorry. I am sorry that I am not aiding you.

2/21/25 Max:
I'm sorry that we might see an increase in coastal oil drilling in the US based on recent executive orders. I mourn for you, because of both the chemical effects of the pollutants that will be emitted when that oil is burned, and because of the ecological disruption that may accompany further enhancement of oil drilling infrastructure on US coastlines.

2/28/25 Becs:
I need more of you to burn as a glaze for ceramics. I am sorry that I seek out your corpse for my own artistic motivations.

2/28/25 Max:

Dear *Macrocystis pyrifera*, I'm sorry - I realized that you are an ingredient in the hair conditioner I bought last week. I have no idea why - maybe your many diverse polysaccharides can adhere and absorb into porous hair? Or maybe they are just being used as filler - a relatively cheap substance that makes the product more gelatinous and texturally pleasing from a human perspective. If the reality is closer to the former, thank you. If closer to the latter, I'm sorry - I don't want you to be used for cheap marketing tricks.

3/14/25 Becs:
I am sorry. I am sorry for ignoring you. I am sorry that it is harder to focus on you. I really am sorry.

3/14/25 Max:
Dear kelp, I wish I could understand directly 'what is important' to you. I know what I think is important for/about you: when I think of kelp, I imagine blades waving under cold waters and molecules diffusing through membranes. But I am certain that other processes exist, which are vitally important and also completely off my radar. What's going on in your stipe? Your holdfast? I have no idea. I'm sorry that my image of you will always be mediated by my (human) notions of 'what is important' about you.

5/19/25 Becs:
I am sorry that sometimes I think that I still don't know you that well. I spend so much time thinking about you. I have spent a good amount of time with you in person as well, and yet so much of you feels unknowable. I am working on it and yet; I am still sorry. I think I might love you, but how is that possible when I don't know you?

6/19/25 Max:
Dear Nereo, it was wonderful to see you this past week. You seem to be doing well on the outer coast - your blades were so clean and huge and healthy! I'm sorry I didn't stay long. I feel kind of like a bad daughter: swooping in for three days to

get what I need, then jetting off again to the next thing. I think about my friends' field work a lot, and how their way of getting to know their subjects involves extended observations (up to 3-4 months total) where they come back every few weeks to check on their organisms. I'm sad that my work is more about taking samples of you and removing them from their natural context, than it is about sitting with you quietly for a while.

6/19/25 Becs:
Hi kelp. I am sorry for not thinking about you more lately. I have spent a lot of time near or in the water. I swam in the Adriatic Sea off Croatia. I snorkeled off the coast of Nice and looked for you, but I mostly saw eelgrass. I might have seen you too, but it was hard to ID and I only had the goggles, not an actual snorkel. I kept taking a big breath and then diving underneath the water until I could feel my lungs burning. I am sorry that I keep looking for you, and don't seem to get that much closer. I'll be at Lake Michigan Friday for a bonfire and again, I won't be near you. I am sorry.

8/31/2025 Becs:
Hi kelp, specifically bull kelp, hi friend. I am sorry for any harm that comes to you from the process of publishing this book. I hope the benefit to you outweighs the downsides of our material usage, but I cannot know. Sorry. I hope to talk with you again soon and see how you are doing.

04/01/2026 Max:
Dear kelp (dead and alive), This week I'm thinking about 'ghost lineages' - the extinct, upsampled, and otherwise unknown branches of the tree of life. The organisms in these lineages exist, or existed at one point in time, but are not directly observed. I'm sorry to your ghosts. It's difficult to know them, and to measure their impacts on the world we see today. But I think as humans we are well-practiced in imagining life shaped by dead ancestors and phantom forces. I'll look for your hopeful sister lineages as they emerge in the future.

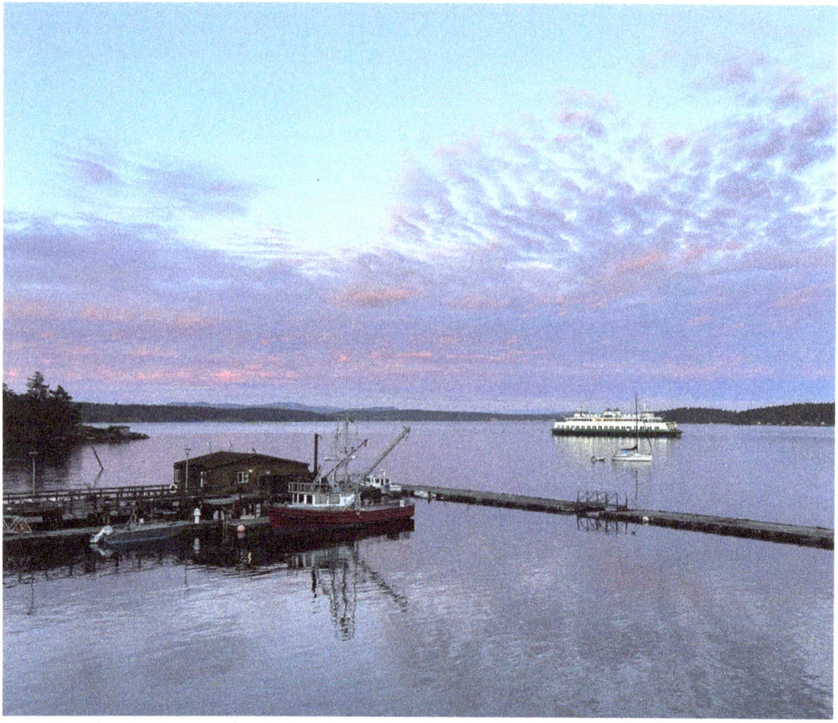
Image 53. Dock at Friday Harbor Labs

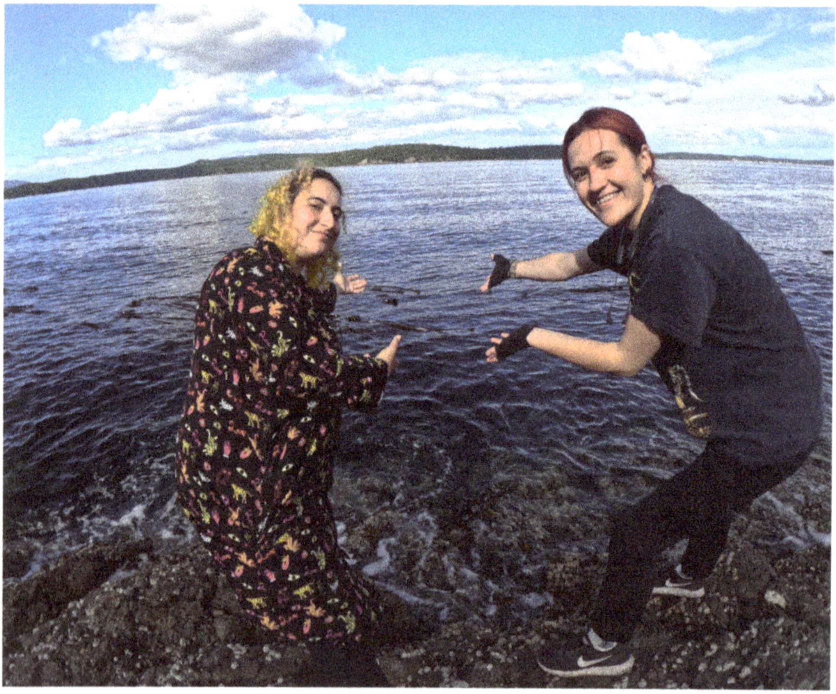
Image 54. Becs and Max on a mission to take video footage of bull kelp.

Acknowledgements

We gratefully acknowledge many friends, colleagues, loved ones, and strangers who contributed to the creation of this book.

We appreciate the support provided by the Arts Science and Culture Initiative (ASCI) through the University of Chicago and School of the Art Institute of Chicago.

We thank the Makah Tribe for their stewardship of coastal ecosystems on the Olympic Peninsula, and for use of seawater samples collected on their land.

We thank Felipe Alberto and the Puget Sound Restoration Fund for all their help and advice regarding kelp husbandry.

Max:

Thank you to Sean Hoban and Sara Black for your mentorship in the initial stages of the project. Sean, your thoughtful perspective on the topic of human responsibility to nonhumans was deeply cherished.

Thank you to Cathy Pfister for use of your lab facilities. Thanks to members of the Pfister and Thornton labs for your time and shared conversations. You all helped me think across scales, systems, and contexts via many direct and indirect interactions over the course of several years.

Thank you to my family for listening to me talk about kelp from the beginning, and for your love, support, and suggestions related to this project.

Thank you to Becs' family for your time and attention. Thanks for buying kelp pickles for Becs to taste, and grammar-checking our entire manuscript multiple times.

Thank you to Nichole Nomura for the incredible care you showed toward this project. Thanks for sharing your expertise related to publishing strategies, and to the place of this work in the broader field of Science and Technology Studies.

Becs:

I want to thank my parents, Tamara Goldman and Marc Epstein, for their tireless and exacting editing as well as for the incredible love and support they give me. I thank my advisors Sara Black, Matthew Goulish, and Andrew Yang who helped me dream big and think through complex ideas. I also want to acknowledge the ways Richard Pell and Bob Bingham shaped the artist I have become.

Most of the microscopy images are thanks to my time as an Artist in Residence at Friday Harbor Labs. I specifically want to thank the entire star fish lab as well as Tom Mumford, Cricket Keating, Deborah Massachi, and Marjorie Wonham.

Thank you to my loved ones; Mina Simon, Marilyn Walker, Kate Epstein, and so many others. Thank you to Max; I am so proud of what we have done together. And finally, thank you to bullkelp and all kelps; you have changed my way of thinking forever.

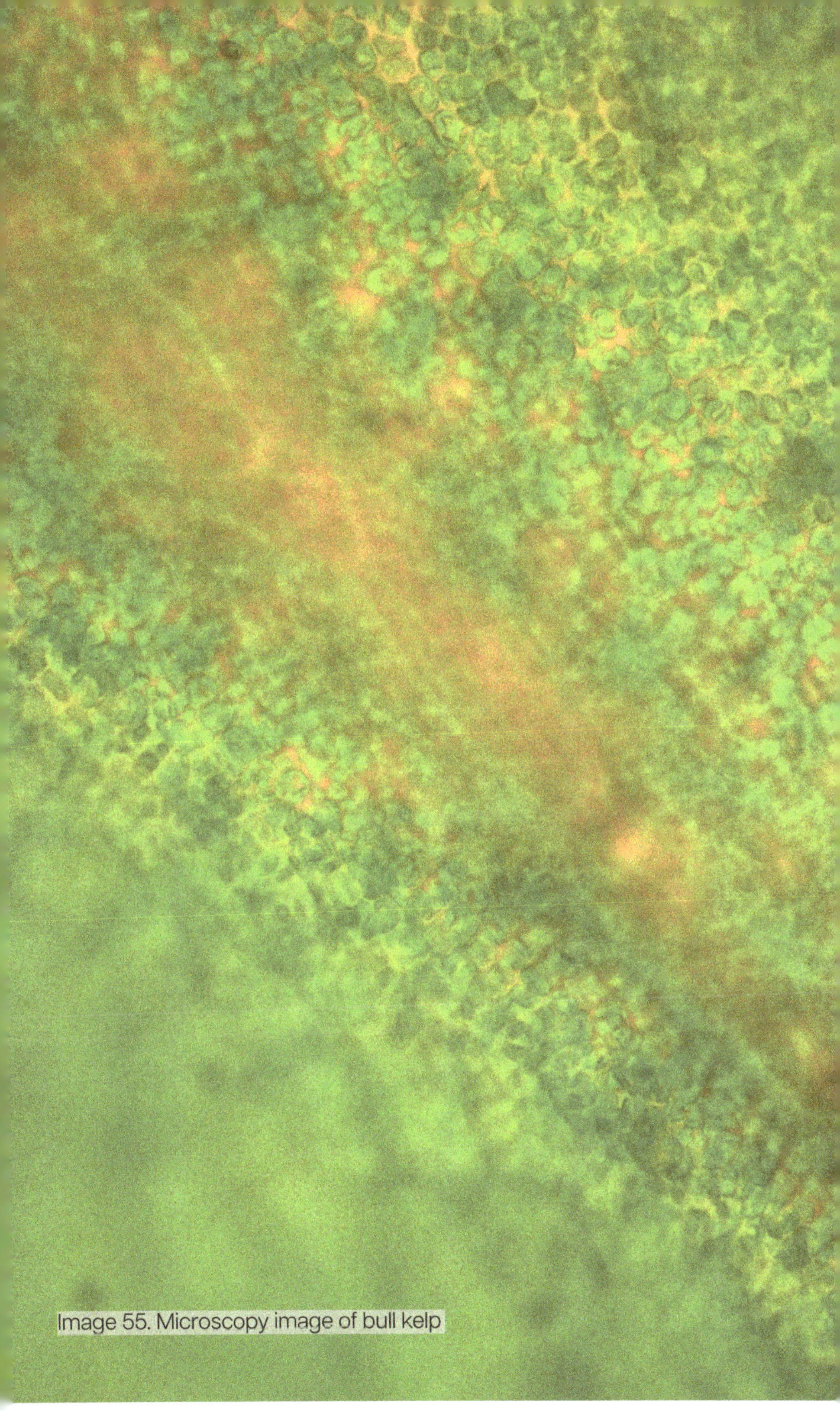

Image 55. Microscopy image of bull kelp

Bibliography

Ballew, Matthew, Jennifer Carman, Marija Verner, Seth Rosenthal, Edward Maibach, John Kotcher, and Anthony Leiserowitz. 2024. "The Attitude-Behavior Gap on Climate Action: How Can It Be Bridged?" Yale Program on Climate Change Communication (blog). March 21, 2024. https://climatecommunication.yale.edu/publications/attitude-behavior gap/.

Bennett, Jane. 2010. Vibrant Matter: A Political Ecology of Things. Duke University Press. https://doi.org/10.2307/j.ctv111jh6w.

Berry, Helen D., Thomas F. Mumford, Bart Christiaen, Pete Dowty, Max Calloway, Lisa Ferrier, Eric E. Grossman, and Nathan R. VanArendonk. 2021. "Long-Term Changes in Kelp Forests in an Inner Basin of the Salish Sea." PLOS ONE 16 (2): e0229703. https://doi.org/10.1371/journal.pone.0229703.

Black Press Media staff. May 29, 2008."Haida Language Play Premieres This Weekend," Haida Gwaii Observer, https://www.haidagwaiiobserver.com/news/haida-language-play-premieres-this-weekend-6329102.

Bringloe, Trevor T., Samuel Starko, Rachael M. Wade, Christophe Vieira, Hiroshi Kawai, Olivier De Clerck, J. Mark Cock, et al. 2020. "Phylogeny and Evolution of the Brown Algae." Critical Reviews in Plant Sciences 39 (4): 281–321. https://doi.org/10.1080/07352689.2020.1787679.

Busby, Cathy. 2005. SORRY. https://cathybusby.ca/publications/sorry.

Cohen, Ed. 2008. "A Body Worth Having?: Or, A System of Natural Governance." Theory, Culture & Society 25 (3): 103–29. https://doi.org/10.1177/0263276408090660. Cook,

Carly N., and Carla M. Sgrò. 2017. "Aligning science and policy to achieve evolutionarily enlightened conservation." Conservation Biology 31 (3): 501–12. https://doi.org/10.1111/cobi.12863.

Dablander, Fabian, Maien S. M. Sachisthal, Viktoria Cologna, Noel Strahm, Anna Bosshard, Nana-Maria Grüning, Alison J. K. Green, Cameron Brick, Adam R. Aron, and Jonas M. B. Haslbeck. 2024. "Climate Change Engagement of Scientists." Nature Climate Change 14 (10): 1033–39. https://doi.org/10.1038/s41558-024-02091-2.

Danto, Arthur C. 2013. What Art Is. Yale University Press. https://www.jstor.org/stable/j.ctt32bfhx.

Davies, Simon.June 2008. "Sinxii'gangu (Sounding Gambling Sticks).," Haida Laas: Newsletter of the Haida Nation.

Ferrando, Francesca. 2014. "'THE BODY' IN POST- AND TRANSHUMANISM - Printed Version." Ranisch, R. & Sorgner, S. L. (2014) Post- and Transhumanism: An Introduction. S.L. Peter Lang Publisher. Pages: 213-226. https://www.academia.edu/10395183/_THE_BODY_IN_POST_AND_TRANSHUMANIS M_Printed_Version.

Flesche, Francis La. 1999. The Osage and the Invisible World: From the Works of Francis La Flesche. University of Oklahoma Press.

Fletcher, Michael-Shawn, Rebecca Hamilton, Wolfram Dressler, and Lisa Palmer. 2021. "Indigenous Knowledge and the Shackles of Wilderness." Proceedings of the National Academy of Sciences 118 (40): e2022218118. https://doi.org/10.1073/pnas.2022218118.

Geib, Claudia. 2022. "How 'Wilderness' Was Invented Without Indigenous Peoples." SAPIENS, February 3. https://www.sapiens.org/culture/pristine-wilderness-conservation/.

Glatthaar, Trisha Corliss, "Tom Price (c.1860-1927) : The Art and Style of a Haida Artist." (University of British Columbia, 1970), https://doi.org/10.14288/1.0102264.

Haraway, Donna. 2006. "A Cyborg Manifesto: Science, Technology, and Socialist-Feminism in the Late 20th Century." In The International Handbook of Virtual Learning Environments, edited by Joel Weiss, Jason Nolan, Jeremy Hunsinger, and Peter Trifonas, 117–58. Dordrecht: Springer Netherlands. https://doi.org/10.1007/978-1-4020-3803-7_4.

Haraway, Donna J. 2016. Staying with the Trouble: Making Kin in the Chthulucene. Experimental Futures. Durham, NC: Duke

University Press.
Hobbes, Thomas. 2002. Leviathan. Broadview Press.
hooks, bell. 2004. The Will to Change. https://www.simonandschuster.com/books/The-Will-to Change/bell-hooks/9780743456081.
Karam-Gemael, Manoela, Rafael Loyola, Jerry Penha, and Thiago Izzo. 2018. "Poor Alignment of Priorities between Scientists and Policymakers Highlights the Need for Evidence Informed Conservation in Brazil." Perspectives in Ecology and Conservation 16 (3): 125– 32. https://doi.org/10.1016/j.pecon.2018.06.002.
Kimmerer, Robin Wall. 2013. Braiding Sweetgrass: Indigenous Wisdom, Scientific Knowledge and the Teachings of Plants. First paperback edition. Minneapolis, Minn: Milkweed Editions.
Krumhansl, Kira A., Daniel K. Okamoto, Andrew Rassweiler, Mark Novak, John J. Bolton, Kyle C. Cavanaugh, Sean D. Connell, et al. 2016. "Global Patterns of Kelp Forest Change over the Past Half-Century." Proceedings of the National Academy of Sciences 113 (48): 13785–90. https://doi.org/10.1073/pnas.1606102113.
Latour, Bruno. 2005. Reassembling the Social: An Introduction to Actor-Network-Theory. Clarendon Lectures in Management Studies. Oxford; New York: Oxford University Press.

Lippard, Lucy R., Rosalind Krauss, Mel Bochner, Dan Graham, Robert Smithson, Susanna Singer, John Hogan, and Sol Lewitt. 2012. Sol LeWitt. Edited by Béatrice Gross. Metz: JRP | Ringier.
Locke, John. Second Treatise of Government. Digitized by Dave Gowan. Edited, with An Introduction, by C.B. McPherson. Hackett Publishing Company, 1980. https://www.gutenberg.org/files/7370/7370-h/7370-h.htm.
Naar, Nicole. n.d. "APPENDIX B: The Cultural Importance of Kelp for Pacific Northwest Tribes."
Pierotti, Raymond. 2012. Indigenous Knowledge, Ecology, and Evolutionary Biology. Routledge.
Riley, Duke. 2022. "DEATH TO THE LIVING, Long Live Trash · Brooklyn Museum." 2022. https://www.brooklynmuseum.org/exhibitions/duke_riley.
Rogers-Bennett, L., and C. A. Catton. 2019. "Marine Heat Wave

and Multiple Stressors Tip Bull Kelp Forest to Sea Urchin Barrens." Scientific Reports 9 (1): 15050. https://doi.org/10.1038/s41598-019-51114-y.
Swanton, John Reed. 1905. Haida Texts and Myths, Skidegate Dialect, with Internet Archive (Washington : G.P.O.), http://archive.org/details/haidatextsmythss01swan.
Tsing, Anna Lowenhaupt. 2015. The Mushroom at the End of the World: On the Possibility of Life in Capitalist Ruins. (144) Princeton University Press. https://doi.org/10.2307/j.ctvc77bcc.
Whyte, Kyle. 2017. "Indigenous Climate Change Studies: Indigenizing Futures, Decolonizing the Anthropocene." English Language Notes 55 (1–2): 153–62. https://doi.org/10.1215/00138282-55.1-2.153.
Yoon, Hwan Su, Jeremiah D. Hackett, Claudia Ciniglia, Gabriele Pinto, and Debashish Bhattacharya. 2004. "A Molecular Timeline for the Origin of Photosynthetic Eukaryotes." Molecular Biology and Evolution 21 (5): 809–18. https://doi.org/10.1093/molbev/msh075.

Figure Descriptions

Figure 1 (Page 48)

Life cycle of the kelp Nereocystis luetkeana. Adapted by Max and Becs from a lecture by Dr. Brooke Weigel

Figure 2 (Page 50)

Experimental design for the growth substrates. We made clay discs, sized to fit into 24-well plate wells. Each disc was made using one of three textures, chemical compositions, and firing temperatures. In total, we created 135 substrates. These were distributed across 9 plates, and each plate represented a different combination of firing temperature and silica content (3 temperature levels x 3 silica levels = 9). Each plate contained 5 replicates of each texture level. The substrates were placed in the wells by hand, and a standard volume of male and female kelp gametophytes was pipetted into each well. Kelp were allowed to grow for three to four weeks

Figure 3 (Page 64)

Normalized count of sporophytes observed on substrates, separated by substrate silica content

Figure 4 (Page 65)

Normalized count of sporophytes observed on substrates, separated by substrate surface texture

Figure 5 (Page 66)

Normalized count of sporophytes observed on substrates, separated by substrate firing temperature

Figure 6 (Page 67)

Distribution of normalized substrate counts across all treatments. The distribution of our data was non-normal, leading us to choose the Kruskal-Wallis Rank Sum Test to test for significance of effect. A. Histogram of counts. B. Results of Kruskal-Wallis test.

Figure 7 (Page 72, 73)

Distribution of normalized substrate counts, separated by counter. A. Distribution of sporophytes counted on substrates, by counter. B. Distribution of sporophytes counted in control wells, by counter.

Image Descriptions

Images taken by Max and Becs unless otherwise stated

Image 1

Bull kelp (Nereocystis luetkeana) near the shore of San Juan Island.

Image 2

Compilation of images of juvenile bull kelp taken at University of Chicago and a tree on San Juan Island. The bull kelp was grown in laboratory conditions and imaged on a light microscope as a gametophyte (see Figure 1).

Image 3

Juvenile sunflower sea star at Friday Harbor Laboratory, University of Washington. The sea star baby was under 3 mm in size.

Image 4

Intertidal assemblage, image taken at Kydaka Point in Sekiu, Washington. Urchins, algae and mussels nestle among the rocks, along with other marine invertebrates. A diversity of life strategies and ecological niches is represented in the taxa pictured.

Image 5

Kelp canopy near the shore of San Juan Island. Mixed brown algal species are represented: the majority of individuals pictured are bull kelp; on the right-hand side of the image, the blades of a closely related kelp (genus Alaria) intertwine; tufts of 'rockweed' (genus: Fucus, a close relative of kelp) are interspersed.

Image 6

Interpretation of bull kelp blade. This blade was one of the first sporophytes observed by Becs and Max in their experiment.

Image 7

Photo of Becs' thesis, photographer Eugene Tang

Image 8

Max holding a bull kelp at Hobuck Beach on the Makah Reservation (Neah Bay, Washington)

Image 9

Becs wrapped in adult bull kelp at Friday Harbor Laboratory.

Image 10

Max with adult bull kelp at Friday Harbor Laboratory. Scale bar = 5 feet 5 ¾ inches

Image 11

Blossoming Moment, oil painting, 2018, Becs Epstein.

Image 12

Dissecting scope at University of Chicago.

Image 13

Macrocystis blades floating at the surface of the water near Bullman Beach, Washington

Image 14

Microscopy image of various moss taken by Becs at Friday Harbor Labs

Image 15

Microscopy image of fabric taken by Becs at Friday Harbor Labs

Image 16

Artwork, *Momento mori*, Acrylic and marker on canvas, 2024, 11"x14," by Max Bogan

Image 17

Becs' art piece created for a Western Tanager she found dead at Friday Harbor Labs in 2024

Image 18

Max and lab mate Daniela Pierro found, held, and sat still with multiple brown algal species at Kydaka Point, WA. The pictured algae is a member of the order Fucales, sister to the order Laminariales (true kelps)

Image 19

Artwork, Apology to Roses from performance *I'm, We're, Sorry*, Screenprint and heatpress on fabric. Heatpressed apologies are from a community workshop, 2022, by Becs Epstein

Image 20

Max spitting on bull kelp to share their mouth's biome with it

Image 21

After Max spitting on bull kelp to share their mouth's biome with it

Image 22

Bull kelp that floated towards Becs at South Beach on San Juan Island. It was her first day seeing patches of health bull kelp. Her new friend, Cricket, took her to this location. Becs pulled off her socks and shoes and rolled up her pants to run into the freezing water to grab this stalk. It seemed to not drift but magnetically head in becs' direction

Image 23

Artwork, *Apology Calendars*, Screen-printed and heat-pressed fabric, rizo print and digital print, 2022/2023, variable sizes, by Becs Epstein

Image 24

Becs kissing bull kelp on San Juan Island

Image 25

Representation of a clay substrate used in our experiment.

Image 26

Photo taken just before mixing the 'very high silica' clay body. Base clay recipe poured over a bed of silica powder

Image 27

Kelp gametophytes, observed under a light microscope.

Image 28

Kelp sporophytes, observed under a light microscope and identified by the emergence of a blade

Image 29

Kelp sporophyte, observed under a light microscope.

Image 30

All three clay bodies prepared in slurry form

Image 31

Photo taken while we pinched the clay substrates to create a 'handle' for lifting and moving

Image 32

Bisque fired substrates on firing trays

Image 33

Photos of a plate, just before UV sterilization. NS = no silica added; SLT = super low temperature. Other silica contents were denoted as MS (medium silica) or HS (high silica); other firing temperatures were denoted as LT (low temperature) or HT (high temperature)

Image 34

A plate loaded with substrates being sterilized by UV light on April 9th, 2023

Image 35

Max pipetting gametophytes over a plate

Images 36-37

Close up of juvenile kelp in control wells. We cannot say for sure whether these were gametophytes or sporophytes. Observed under a light microscope on May 5th, 2023

Images 38-39

Close up of juvenile kelp attached to a substrate. We cannot say for sure whether these were gametophytes or sporophytes. Observed under a light microscope on May 5th, 2023

Image 40

Most of a 'smooth' substrate surface, showing juvenile kelp attached. Observed under a light microscope on May 8th, 2023

Image 41

Close up of a kelp sporophyte. Observed under a light microscope on May 8th, 2023

Image 42

Close up of a kelp sporophyte. Observed under a light microscope on May 15th, 2023

Image 43

Artwork, *A Vision*, oil painting, 2022, 20"x36" by Becs Epstein

Image 44

Artwork, *Nereocystis Postels*, by Josie Iselin

Image 45

Artwork, *Knotted Light*, by Rochelle Voyles

Image 46

Sketch by Max

Image 47-48

Artwork, *Beneath the Tide*, Laser-engraved ceramic, 2023, 2'x3'x1.5" by Becs Epstein

Image 49

Sketch by Max, *Stably Holding tension and passing along vibrations to the network*

Image 50

Slightly desiccated young bull kelp individual, washed ashore near Friday Harbor Labs

Image 51

Microscopy image of bull kelp taken by Becs at Friday Harbor Labs

Image 52

Burning seaweed (mazzaella) collected by Becs at Friday Harbor Lab to make ash glaze

Image 53

Dock at Friday Harbor Labs, observed from a small hill on the campus

Image 54

Becs and Max on a mission to take video footage of bull kelp near Friday Harbor Labs

Image 55

Microscopy image of bull kelp taken by Becs at Friday Harbor Labs

Table Descriptions

Table 1 (Page 51)

Base recipe for the clay body used to make the substrates.

Table 2 (Page 52)

Table of variables, with explanations of the levels of variation. The different firing temperatures were achieved by changing settings on the kiln. The different textures were achieved by pressing different materials into the surface of the clay before firing. The different silica concentrations (weight/weight) were achieved by adding different quantities of silica powder to the base Red Earthenware recipe. The base is 42.5% silica, and this served as our lowest silica level. To achieve 55% silica by weight, we added 1.33 lb of pure silica powder to 4.8 lb of the base. To achieve 65% silica by weight, we added 3.09 lb pure silica powder to 4.8 lb of the base. Max calculated the amounts of silica to add, and Becs mixed the powder into the base by hand

Table 3 (Page 53)

Abbreviations used in naming plates

www.ingramcontent.com/pod-product-compliance
Lightning Source LLC
LaVergne TN
LVHW051923051126
851857LV00021B/2109
* 9 7 9 8 9 9 6 5 5 3 6 0 0 *